A THOUSAND TRAILS HOME

LIVING WITH CARIBOU

SETH KANTNER

**MOUNTAINEERS
BOOKS**

MOUNTAINEERS BOOKS is dedicated to the exploration, preservation, and enjoyment of outdoor and wilderness areas.

1001 SW Klickitat Way, Suite 201, Seattle, WA 98134
800-553-4453, www.mountaineersbooks.org

Printed in China
Distributed in the United Kingdom by Cordee, www.cordee.co.uk
23 22 21   1 2 3 4 5

Copyeditor: Ali Shaw
Design and layout: Jen Grable
Cartographer: Lohnes+Wright
All photographs by the author unless credited otherwise
Front cover photograph: *Caribou move south on new sea ice in late fall.*
Back cover photograph: *A cow and calf stay close together as they travel south on the tundra.*
Opening portfolio: Page 1: *Caribou follow ancient trails across the mountains of the Brooks Range as the herd begins to aggregate.* Pages 2–3: *Caribou flee from humans on snowmobiles on the Baldwin Peninsula. Bubbles of methane released by melting permafrost are trapped under lake ice.* Page 4: *Shortly after giving birth cows lead their newborns across the Kokolik River.* Page 5, top: *A cow and calf on the fall migration.* Page 5, bottom: *Caribou after crossing the Kobuk River on the migration north to their calving grounds.* Frontispiece: *After eating very little during the rut, bulls resume feeding along the fall migration.*

Some portions of this book originally appeared in different form in other publications, including *Caribou Trails*, *Adventure Journal*, *Arctic Voices* by Subhankar Banerjee, the *Anchorage Daily News*, and other regional newspapers.

Library of Congress record is available at https://lccn.loc.gov/2021008622, and ebook record is available at https://lccn.loc.gov/2021008623.

Mountaineers Books titles may be purchased for corporate, educational, or other promotional sales, and our authors are available for a wide range of events. For information on special discounts or booking an author, contact our customer service at 800-553-4453 or mbooks@mountaineersbooks.org.

Printed on FSC®-certified materials

ISBN (hardcover): 978-1-59485-970-0
ISBN (paperback): 978-1-59485-972-4
ISBN (ebook): 978-1-59485-971-7

An independent nonprofit publisher since 1960

For Stacey—
who came north, somewhat accidentally,
and a far greater distance than either of us realized,
to live this life beside me, one with more caribou, bears, wolves,
and wildness than a young Boston woman may have ever dreamed

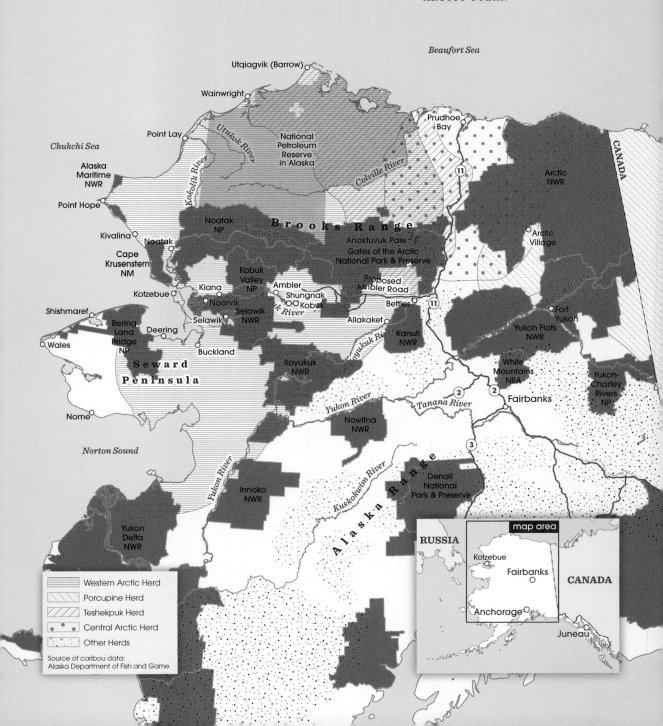

CONTENTS

The arrival of ice and fresh snow overnight alters the landscape the animals must traverse.

CARIBOU IN THE NIGHT

Barefoot in the darkness on thin fall snow, I stop walking. My feet will be freezing soon, and I'm only here to listen. Forty steps behind is the small yellow glow of lamplight coming from the window in my sod home. Inside, I have hardwood steaming for runners for a dogsled, and I need to get back in to make sure the water doesn't boil dry.

Above the branches and brush, the stars are sharp and a thin fingernail moon offers little luminance. From the north comes a cold breeze, hardly more than a stir. The aurora is weak too, faint green smoke up against the stars, and not enough to light the night. Ice pans far out in the river make soft roars as they collide with heaped fast ice, tinkle, and then spin silently on their way west.

My feet are feeling the cold. South, across the dark river, float the comments of a longtime companion, a great horned owl. *Whooo. Whooo. Whoo-whoo. Whooo.* From down the ridge comes the rattle of chains, and a few whimpers and whines—my dog team, pacing in circles, restless in the rich-smelling darkness. They, too, are likely holding freezing pads up off the snow, one at a time to thaw, while they focus their ears north. For once I'm not out checking on the team. I'm not

Snow, ice, and the winds of winter begin to transform the land.

trying to dissuade a middle-of-the-night grizzly bear from borrowing my meat. I'm only listening.

From the north, the night slowly fills with soft sounds: many feet moving through snow and the low wiry brush of frozen tundra, the click of hoof tendons, an occasional grunt, the clatter of antlers sweeping dwarf birch and alder branches. Caribou are passing in the night.

Big and small herds have come through for nearly two months now. It's only the last three days, since the river began running heavier ice, that this uninterrupted line of animals has been marching east. In truth, caribou have been passing my entire life; the land is veined with their ancient trails. But something is different tonight. There's something big and dark and wild about standing barefoot on thin snow and frozen ground, hearing thousands of animals traveling through, and not being able to see even one. It's exciting and humbling, shivery, and on the edge of scary—as if some huge parallel nation is on the move out under the starlight.

Now, my feet have frozen patches. I'm suddenly cognizant of the pinch and burn, and I hurry back toward the small yellow glow, a lone light in this world of darkness. Inside, on my family's old bearskin couch, I sit and feel the twin aches: of skin thawing and of love for my home, this land of caribou.

FOLLOWING PAGES: *Glowing red blueberry bushes brighten the tundra with splashes of color in early autumn.*

On an evening in mid- to late August, there comes a chill to the air. Along the shores of the river, the grasses and willows have lost the green of summer and are tinged with yellow. Across the tundra, distant ponds glow gold as the sun goes down, further than it has since April, hinting of winter returning. To the northwest the mountains are blue-black triangles below the electric orange, green, and blue of the Arctic sky. Shadows line the shores of lakes; the water mirrors the sky, like molten metal, with the black fingers of spruce trees reaching out across the surface. Overhead the first stars scrape through the blackness after the sunny nights of summer, and in the falling twilight, the only thing bigger than this vast land is the silence. Silence as hushed and huge as space.

Finally, one small sound floats from the nearby grasses—a rustling of voles in the leaves, there and gone. After a minute comes another noise: the flutter of wings and the soft purr of feet on water as a family of mergansers lifts off the river, vanishing into the coming night. East, over the hills, an unidentified raptor shrills, a single faint call. And then from the west floats an unmistakable sound, the splash of caribou hooves crossing shallows. Tonight, in these few moments, fall has arrived—infused with energy and rush, hunger, and the necessity of gathering food.

What is ending—the green drone of summer, leaves and mosquitoes, warmth and rain—always feels like an intrusion in the Arctic. Only in autumn does the land begin the swing back to being this home of ice and snow that we know best. As the temperature and taste of the air change, every living thing recognizes the changes: the beetles crawling on the mud, the equisetums dying back to earth, the no-see-ums rising like dust in the warm afternoon air, the high-bush cranberries dropping their red leaves, the shrews tunneling in the grass, the bears digging dens, even the birch roots down in the ground stashing nutrients to be carried skyward by sap in the spring—everything knows winter is coming.

Now each day the tundra glows brighter, tinted with more vibrant colors, until the land is lit in a burgundy carpet rolling toward the mountains, with splashes of yellow birches painted on ridges, and far away a fiery orange tree on that knoll, and there another, off against the horizon, standing alone before the deep-blue mountains. Other trees already have bare branches, boney stick fingers, leafless, all packed and prepared for winter.

Across the tundra, winding into the land, the sloughs and creeks are pale yellow squiggles, marked by willows still clinging to their leaves, and higher up the mountain slopes, green valleys of alder taper to red where bearberry cling close to the ground, their leaves as brilliant as fresh blood dripping from a wounded caribou's lungs.

Along the borders of the tundra, behind bands of tall spruce, thickets of dwarf birches hold their million leaves up to the blue sky like tiny seashells painted red, yellow, and orange. Wiry blueberry bushes tangle footsteps while their berries invite hungry mouths. The blueberries hang plump, beautiful chalky blue, sweet and irresistible under purple and red leaves ready to fall and flutter free. From knobby tussocks, cranberries dangle ruby pearls and green leaves like miniature upside-down spoons. Crowberry tendrils mat the ground with berries black as mink eyes, juicy and seedy. In between and underfoot, random and everywhere, Labrador tea grows enthusiastic and hardy, its sharp scent mixing with all the other pungent perfumes that rise from this concentrated mat of vegetation to form the unforgettable fragrance of autumn on the tundra.

ON THE VAST SWEEP OF OPEN TUNDRA, BANDS OF MIGRATING CARIBOU DOT THE DISTANCE like quartz outcroppings. And eventually, from closer, come sweeping-branch sounds from the brush—bull caribou forcing their huge antlers through thickets— and a small herd files out onto a tundra meadow. They glance around, checking for danger, and pause to feed and to make decisions.

Today they choose to lie down on the sweetly scented tussocks, to rest during the heat of the afternoon. In no rush, they await a leader to step forward, to choose the next fork in the trail, a trail that leads perpetually to more trails—endless choices in the path ahead, a thousand forks one day, a thousand more the next— carved and cut into this land down through the ages by the hooves of a billion ancestors, leaving the earth itself altered by their journeys.

Now, in late August, the animals are still dark and nondescript. Their hair is short, dark brown—almost black—with white trim. Their diet is transitioning from a summer of willow leaves, sedges, flowering tundra plants, mushrooms, and

Cows and calves lead the migration, usually with young bulls following.

other fresh growth to an autumn diet of lichens, grasses, willow tips, and other vegetation. As the season turns to autumn, the flavor of their meat changes quickly from a slightly boggy flavor to a cleaner taste. Bulls feed on dwarf birch, stripping the leaves, instinctively ingesting the calcium-rich leaves necessary for hardening off their antlers.

Occasionally, the animals pause from eating to pick fights with small innocent green spruce in their path. Caribou are the only deer species in which both the males and females have antlers, and these mock spruce battles strip shreds of velvet off, leaving them red and bloody and soon burnishing to brown. Fighting with spruce trees also provides the adult bulls their first practice sparring in almost a year, since the last rut. And who knows, maybe these fine bulls are playing out fantasies of overpowering opponents, winning fights, and mating in the warm afternoon. And now, sure enough—sudden sharp clattering sounds carry from a nearby rise. Two large bulls have their heads down and their antlers sparring, for a few seconds almost serious, and then as quickly as it began, the duel is done. This is still practice after all, and fairly congenial. The rut hasn't yet begun.

THE HERD BEGINS TO MOVE, SLOWLY, INDECISIVELY, BUT SOON RAPIDLY, A STEADY POURING away of animals, dropping over a ridge and disappearing down into alders. Finally, below, a quarter mile away, the leaders appear again along the bank of the river. An adult female wades into the water. In the distance comes the faint grunts and bleated "Ert! Ert!" of cow caribou calling their calves and the calves answering in their more plaintive "Eerrt! Eerrt!" In the sky, ravens caw, and somewhere high overhead and unseen, sandhill cranes circle, dots in the blue, conversing in their rusty-hinge-like calls.

Another raven passes overhead, and then two more, scanning the tundra and brush and shoreline, their wings panting sharp breaths out of the air. Randomly, in midflight, their black forms plummet, falling toward the ground, only to roll back into flight and smoothly fly on, playing yet always watching, patient and missing nothing, cawing their commentary of caribou across the miles. Occasionally one will land in a treetop, perch and make suggestions, and soon fly on toward other opportunities. Far away on the tundra, dark dots will be black bears—but no!—those too are ravens, congregated to eat blueberries while they wait for a successful hunter. Or, wait, maybe it's a caribou they are feasting on, dead down in the tussocks, left by a grizzly bear. Sure enough, more ravens are arriving, flocking to enjoy their favorite feast—caribou eyes and tongues and intestines, sweet and fat, fresh and warm.

OVERNIGHT THE BEAUTIFUL BURGUNDY TUNDRA FADES TO BROWN. A FEW MORE DAYS AND the land is almost gray in its acquiescence to winter's coming. The north wind blows relentlessly now. In the morning, ice is on the ponds, with windy blue waves lapping in narrow slits not yet frozen. Afternoon sunlight is the color of straw. Meanwhile, still coming, are the endless caribou herds, their presence making this fleeting fall bearable, all part of life here.

The caribou change almost as fast as the land—moving in the opposite direction along the plane of beauty. In the time it takes birch leaves to turn yellow, fade, and fall, the animals change from drab indistinct dark gray to rich brown, with white bellies and rumps and lines along their lower flanks. Their neck hair grows long and white as snow, almost too quickly to believe. Oestrid flies, a curse in the lives of caribou, for now have receded with summer, and the flies' offspring— torturous warble and nose bot larvae—ride as tiny hitchhikers under the skin and in the sinuses of the animals. The bulls are heavy, burly after a summer of putting on fat—averaging three hundred fifty to four hundred pounds, with thick powerful necks draped with white manes—and holding high their glorious antlers, tall and broad, burnished and hard. The cows are half their size, with antlers half sized, too, and are healthy and plump, with a soft, stressed look in their eyes. The females are the leaders of this nation, and they take turns guiding groups across this dangerous hungry land. At their sides are their calves, still nursing, with small velvet antler spikes and lovely, gentle dark-and-silver faces.

Already the adult males are entering a hormone haze. Soon they will be lost in it—arrogant and proud, careless and stupid, thinking only of mating—and doing little to assist in this major migration to the wintering grounds. The bulls will eat little or nothing, their stomachs will shrink, their penis sheaths will become extended, and they will spar vigorously with other bulls. Many will die of wounds inflicted on each other, and from wounds that slow them enough to be eaten by wolves and bears. Along the riverbanks, lost calves will bleat for their mothers, separated and confused, pacing in fear. On other shores, distraught mothers will call and call, often for days and often in vain. In the willows and tangled thickets along creeks and sloughs, groups will randomly flee from threatening sounds and the scent of wolves or bears. The terrified cows and calves and young bulls will outpace

Lowbush cranberries, or lingonberries, grow on ridges and hills and from tundra tussocks.

Caribou antlers are bright red when the velvet is first stripped away and quickly burnish to brown.

the big bulls with their huge antlers tangled for precious seconds in branches and limbs—time enough for grizzly bears in the last days before fall denning to end these animals' journeys in the cool and pleasant privacy of the deep dark brush.

DYING, THE LAND IS AS ALIVE AS IT EVER GETS, HUGE, SPRAWLING, VAST AS IMAGINATION itself, wild as you could wish, intense with life, beautiful, and every day receding too fast. Amid this bounty and beauty, it is only human to desire it not to end, to wish to somehow slow this passing of time, even for just a day or two. Yet the seasons march on, swift and unstoppable. This relentless cycle of nature is contagious; such concentrated beauty and bounty, placed so directly in the path of what you know is coming—the long gray months of winter—compound an already fierce incitation to gather. Nearly every living thing out here is engaged in one form or another of preparing, and you, too, are drawn to gather all you can before this time is over.

But for just a short while, toward evening, you walk to a high point and stand staring across the tundra to the north. After a few minutes of eyeing the land for animals, instinctively your glance drops to scan the vegetation around your boots for berries. There in the crumbly black dirt of the outcropping, a glint catches your eye—a chip of rock, an ancient castoff. Caribou hooves have unearthed this broken bit of an obsidian spearhead. Or maybe it was a skin scraper. In your palm, the scalloped glass is a sudden whisper from the past, proof that lives and lives and more lives were lived out on this well-traveled ground. All of them exactly as large and hungry and fragile as your own.

You rub the dirt off the stone, peering into the smoky depths of the volcanic glass, examining it for a few moments before pausing, and shrugging, and flipping the fragment back where its tiny message might continue on through the centuries.

Your hands hang at your sides, calloused and scarred from work, stained from crowberries and crusted with blood from cutting fish and meat. You stare out over the tundra, inhaling the fragrance and beauty with your lungs and your skin and your eyes. And out there still they come, walking, plodding, pausing now to graze and gaze around themselves—more and more and more caribou, the thousands of individual animals that make up this mass migration. And like the fall that is fading, it seems nothing can stop this river of caribou flowing from the north. Little else left alive on the ground of this continent can compare, and in the slowly failing twilight, you are a piece of the wild, with countless companions surrounding you in every fold and crevice of the land. The feeling of belonging fills your chest and overflows out across the tundra to the walls of your world, the mountains and the blue edges of the sky.

FOLLOWING PAGES: *The aurora lights up the night sky over the Noatak Flats. Red bearberry leaves and green crowberry needles brighten the hilltops, knolls, and mountain slopes.*

Caribou instinctively fear thin, new fall ice, and extended periods of freezeup can lengthen the time herds must wait to cross rivers and streams.

When winter arrives, the animals begin digging, or cratering, for food and switch to a diet largely made up of lichens, supplemented by dormant sedges and small shrubs.

As alders, dwarf birches, and brush increases, bulls face extra challenges navigating tundra thickets.

In fall grizzly bears feed on salmon, berries, and occasionally caribou—often bulls—whose unwieldy antlers get tangled in willows and alders.

Quickly, he checked the wood supply and stoked the fire. In the flickering light from the flames, the empty cartridges in our hands winked and glowed like gold. Kole and I sniffed at the brassy metallic smell and the leftover gun smoke inside, and the faint oily scent of reloading lubricant on the shoulders of the shells. We blew across the openings, making them whistle, and carefully tapped the shells upside down on the part of our floor that was not dirt but made of round alder pegs pounded into the ground with the tops sawed off square. We'd seen our dad tap brass like this before, to make certain no lint or caribou hair was inside.

Kole took all the empty cartridges and went to the corner and reached up to the workbench, too high up for me. On the bench were hand tools: saws and awls and reloading supplies. Along the back corner was an old brass breadbox and scattered junk: wire and bolts and an old jade ax-head, a few bits of broken chert, and a strip of ancient caribou-antler sled runner. Kole lined the cartridges upright on the edge, like tiny golden soldiers in a row, awaiting the magic that we would help perform on them in the coming night—to bring back their power.

We scampered back up on our bunks to warm our feet on the caribou-hide mattresses. My dad disappeared back out the entrance to head north and retrieve another load of meat with his dog team. Later, when he returned, he brought in wads of *itchaurat* (lacy stomach fat) for my mom to fry with fresh liver. Again, he stepped down into the entrance tunnel, hunched low, and vanished out the caribou-skin flap. When he reappeared, he carried in an armload of firewood and a clump of short, fat front ribs. Finally, he took off his outer garments and caribou mukluks. He whipped the snow out of the fur and hung each item from a peg to dry, and then swept the loose snow down into the entrance to shovel out later. The world seemed at peace now, with the wind and cold and wild land outside, and our family safe inside, in the protective ground, under the insulating snow.

At our small kitchen counter, he quickly cut up the brisket and ribs. Mama put them in the roaster for soup and added seasonings and water, while my father turned to make coffee for himself and Mama. Our little family gathered around the stove to eat liver and *itchaurat* out of the black frying pan. Afterward, my parents sat together on the edge of their bunk with their socked feet pointed at the stove and their hands wrapped around mugs of coffee. Their wool socks were caked with twisted gray caribou hairs, their sweaters sprinkled and twined with the shedding hairs. Stray caribou strands clung to their pants and lay in a fine mesh on the floor and in the corners. My dad sipped from his cup, pulled one off his tongue, unconsciously wiping it on his leg. "Windy out there today," he commented. The weather decided his every day, many decisions, and was part of nearly every conversation.

Kole and I leaned against his legs, our eyes wide and waiting for the tale of his time out on the tundra. "What did you get, Howie?" we asked. "Tell us! Tell us!" (We called him Howie, not Dad; that's what Mama called him, and we boys didn't know any different.)

He grinned down at us. "Cold out there. Chilly north wind again." He said that often, for good reason—the north wind blows relentlessly here. Howie pressed his palm to his cheek and told of long lines of caribou flowing out of the mountains, crossing the new ice on the Hunt and Nuna Rivers, working their way across the flats and up the ridge to eventually disappear on the high tundra, migrating east toward Onion Portage. He had left his dog team out of sight in a draw and walked farther out onto the tundra, the tussocks underfoot stretching on and on to the north, millions of mottled brown-and-white heads, grassy and knobby, drifted with snow on the lee sides, whiskered with the gray shrubbery of dwarf birch, blueberry bushes, and low Labrador tea.

Eventually, Howie came to an overlook—a place he had named the Second Birch Knoll, or the Far Birch Knoll, depending on the conversation—and there he knelt in the snow, waiting, watching the tundra, while his knees grew icy and streaked the snow with red where he had crushed cranberries down in the buried vegetation. The Jade Mountains were five or six miles to the northeast, like burly blue-and-white giants leaning against the back wall of the sky. Below him the country dropped away, rolling across the tundra plain. Into that vast landscape the timbered line of the Hunt River branched and branched again—into five rivers: the Akiak, Hunt, Nekakte, Nuna, and Akillik—and those valleys branched into more and more tributaries, some dark with stands of spruce, others gray with alders and willows, all rising toward the peaks in the north, the Baird Mountains.

My dad waited there. He didn't want to go farther, as that would mean hauling the meat too far. He preferred to let animals come to him. More often than not, he

Howie, Erna, Kole, and I stop on the trail to rest the dogs. (Photo by Don Williams)

hunted within a hundred yards of home. Wisps of cloud stretched white hooks in the pale blue sky. Ravens watched from overhead, dark dots playing on the wind. Wolves were out there, too, somewhere—tracks in the snow proved their patrols here.

On the tundra, windblown snow drifted around his mukluks and sifted into the flaps of his muskrat hat and the eyepieces of his binoculars. His eyes watered from the cold wind and from the emotion, too, of glassing out over so many miles of beautiful wild land and the intensity of the hunt. Howie watched the progress of far-off herds, specks in the distance. Meanwhile, he worried about his dogs—a mile back and now maybe chewing the lines, or fighting, or slipping their harnesses and running off, chasing after the provocative scent of so many passing animals. He grew chilled and tensed his muscles to warm up while waiting. He told himself just a bit longer, just ten more minutes. Later, he told himself five more minutes, and five more after that—whatever that means to a hunter who doesn't own a clock.

Occasionally he brushed aside the drifts to find cranberries. He plucked the frozen red beads from the leaves and snow, sucked them in his lips for something to do to pass the time and for that small encouragement of gathering from the land while waiting on the more vital gathering, meat.

A movement caught his eye. A caribou trotted out of the alders on the south bank of the Nuna. More animals followed. Far in the distance, a line of caribou flowed onto the tundra. The herd worked their way across the flats. My dad couldn't leave now. He checked his rifle barrel for snow. He ate a few more frozen berries. He rubbed his fingers, warming them as he crouched lower behind a large tussock.

Finally, the caribou came briskly up over the ridge. Small bands flowed on both sides of him. He picked out fat ones and watched to make sure the big bulls were feeding still and their stomachs were round, and there in the wind he shot five. When the animals were down, he carefully noted where each had fallen, retrieved his empty cartridges from the snow, and bounded across the tussocks toward the first of two that still struggled. He didn't want to use another bullet and instead grabbed the antlers and wrestled the powerful bull's head, nose down, holding it there with one hand while he plunged his knife in at the base of the skull.

He finished off a second struggling animal, and then began gutting them. When he was done, he left the carcasses whole and cut alder branches to attempt to protect them from ravens, hung his handkerchief to sway in the wind to frighten the birds, and hurried back for his dogs and sled.

THE STOVE CRACKLED. HOWIE LEANED FORWARD, DUMPED THE COLD DREGS OF HIS COFFEE back into the pot, and poured himself a hot cup. Above the tunnel entrance, the light coming through our flapping Visqueen window had gone from the pale of day, to blue, and now gray. Soon it would be black with night.

Mama bent over the kerosene lamp to light the wick. The flame rose and threw shadows under the workbench. On the stove, teakettles sang and the roaster pan simmered with caribou brisket bones. The pan was an enamel roaster, standard in most homes in the villages, used for cooking large portions of caribou. The pelvis or even a whole caribou head—skinned and with the antlers sawed off—would fit inside. Caribou head and caribou pelvis were favorite meals because they contained so much fat. Anything with fat was coveted.

Kole and I hovered close to the warmth of the stove. We were barefoot, our toes red and dusty on the cold dirt floor. The smell of broth made us hungry. The porcelain coating on the roaster was black with shiny white specks, like a dark starry night, and we stared into the blackness as the aroma of soup filled the room.

"Were the bulls fat?" Mama asked. Quickly she added, "Did they smell?"

We glanced up, watching Howie's face, waiting—even as little kids we knew the importance of his answers. Those were the two most important questions a person living off the land could ask about caribou harvested then and there. Bulls would begin rutting in the first week of October, and hormones would flavor their meat. My parents' friends—Eskimo hunters and seamstresses, lifelong experts in all things caribou—referred to caribou in rut as "stink" or "getting stink." The smell was sharp and off; the blood smelled skunky on your hands; cooking the meat filled the house with the odor. Even our sled dogs, who would wolf down almost any meat or fish, frozen or rotten or dried, would balk and sniff suspiciously at rutting meat.

In a month, by the time the rut was over, the adult males would have virtually no fat remaining on their bodies. Their valuable savings—a summer's worth of accrued fat—would be spent, and soon their glorious antlers would be gone, scattered on the trails south. The powerful and glamorous bulls would have transformed into skinny, antlerless animals with rapidly thinning necks and backs becoming knobby with warbles growing under their skin. Transformed, too, from a vital source of food and fat for humans to make it through the winter into gaunt animals themselves facing starvation and the long struggle to survive on the snow-covered tundra.

Most families living along the river and in the nearby villages were similar to ours in their need for all the fat they could gather for themselves and for their dog teams to remain fit and healthy enough to work through the winter. Every local hunter knew the disappointment of hurrying to a downed animal only to see that

Hides hang from pole racks during freezeup, with our log cache and sod igloo in the background.
(Photo by Don Williams)

it was skinny—putting a blade in cautiously, still hopeful, only to find no fat around the penis, no *itchaurat* encircling the stomach, no fat on the back.

A skinny animal meant the same amount of work as a fat one, but with little reward. In the Arctic, in old Eskimo culture, an animal killed for food that had no fat sent feelings of disappointment, failure, and humiliation straight to a hunter's heart. The hunter's eyes immediately lifted to the horizon, searching for more animals, seeking a replacement for this skinny, undesirable creature. Too many skinny animals in a row, and the hurt translated into hunger for entire families. Here, most men had identical careers: hunter. Success was measured in fat and in furs—both requirements for life. Without fat and furs came cold, danger, and even death. Every animal without fat—in every season—was a bleak reminder of the nagging need to stave off hunger with another, more successful kill. There was no substitute for hunter success.

My dad held up his hands for my mom to smell the blood on the backs of his wrists. She sniffed carefully. Finally, she nodded approval. He grinned and said the blood had been so greasy, it was hard to grip his knife handle. He said it reminded

Caribou trails stretch for miles across the land and sea ice.

Howie was done eating, he slipped on his mukluks, preparing to head out on his nightly rounds—to check the dogs, feed them, put away harnesses and shovels and axes, and haul in firewood.

"Is the snow good?" Kole asked. We clapped our hands. "Is it fluffy?" We wiped our fingers on our grubby little shirts. "Can we make ice cream? And are we going to reload tonight?"

Howie grinned. Before he disappeared out the tunnel, he reached up to grasp our long shiny metal spatula where it hung from a nail. The spatula was a simple signal: it meant he would search in the lee of the hill for fluffy snow, fill two buckets, and stash them at the end of the tunnel where the snow wouldn't melt while he did his rounds. That spatula meant snow ice cream.

We heard the dogs hollering—sign that he was feeding them—and then his face reappeared. He brought in an armload of wood, and another, and then an armload of caribou legs to skin. Last, he brought in the buckets of snow. Kneeling on the floor, he reached up for materials for snow ice cream. Into our big stainless steel bowl, he scooped snow, sugar, Darigold powdered milk, and a pinch of salt. He stirred it, allowing Kole and me to take turns adding spatulas of snow. He began to stir faster. When the snow turned to white slush, he added a teaspoon of vanilla and began whipping. Soon the big bowl was filling, rising with white creamy snow ice cream.

"Got your bowls?" he ordered, whipping harder. He plopped big dollops into our dishes. We backed away, gripping the sides, staring greedily, spooning the delicious ice cream into our mouths. We sat on a caribou skin on a wooden Blazo box by the fire, pointing our bare feet at the stove to warm them while we ate and stirred, ate and stirred, until we couldn't eat another spoonful.

Afterward, Mama asked if we were ready for a story. She opened a big brown tattered copy of *The Thousand and One Nights*. Kole and I leaned on both sides of her to listen while she read. Howie filled the lamp, fed the fire, and put the leftover ice cream down in the entrance to freeze. Then he sat on a stump and began to skin the first caribou leg.

Mama read about a boy named Aladdin who was trying to retrieve a magic lamp from a cave. It was dangerous and frightening for him. Howie sharpened his knife on a small round gray whetstone, listening while he worked. He cut and pulled at the legging, smiling occasionally when he found something in the tale humorous. When he was done skinning the last one, he took a meat saw down off a peg and sawed up the leg bones for dinner tomorrow. There was too much for one meal, and he put the extra down in the entrance. We boys watched him cleaning up, meanwhile swallowing Aladdin's adventures with our ears. We started listening less, getting twitchy: Howie was cleaning up, and that meant it was time to reload. As little kids, with the huge wild winter world outside the walls of our igloo, reloading was the closest thing to magic in our daily lives.

On the floor, Howie stacked the legging skins, hair to hair, skin to skin. He placed the stack on a Blazo box to keep the shrews off. He rose and washed his hands in the basin. Finally, he went to the workbench and set up his gunpowder scale. He peered at it, setting the balance bar in place, zeroing the weights by twisting the threaded dial until the marks on the scale lined up. Mama paused; she glanced up. Instantly, we boys scampered over to see what he was doing. Howie chuckled, and he rolled two stumps over so we could stand on them. Our red toes gripped the tops of the stumps. It was nice to be up where it was warm.

The reloading press was dark green, heavy cast metal, and bolted to the bench. Howie screwed a silver metal die into the top and a metal rod with a green handle into the bottom. Carefully, he wiped the shoulders of each brass shell with a tiny rag with a dot of resizing lubricant. He had a small tube of the lubricant; the tube was triangle shaped and made of white plastic. It seemed very small, very finite, invaluable, and we weren't allowed to touch it. As he finished lubricating each brass cartridge, he handed it to me.

The empty brass needed to be pressed in the die, one at a time, to remove the primers and resize the brass. In that moment, carefully sliding it into the curved shell holder, I felt thrill and complete focus. I felt important, on the trail to being a grown man. Kole lifted the handle, I got my fingers out of the way, and the primer popped out the bottom. I watched him, wishing for the day I would be able to lift the handle. Next, Howie lifted the handle even farther—the hard part, where the brass gets resized. This part was serious and scary. Sometimes a cartridge jammed and wouldn't come back out of the die. Then Mama stopped her sewing, and we all kept our eyes on Howie and his struggles to get the brass back out. At those times,

While working on meat, Howie leans his hunting rifle on an old kill from years past.

he would question himself: maybe he'd used too little lubricant or reused that cartridge too many times. Too much lubricant was not good either. It seemed terribly important that he get the damaged cartridge out—as if it meant life or death, just like in our book of fables—as if we and the dogs might starve if he couldn't turn those empty brass chambers back into bullets by morning.

When the brass was resized, Howie put a primer in the spring-loaded holder. Kole lifted the handle up, Howie pressed the holder forward, and Kole pulled the handle down. We did this ten times, for ten rounds, and then Howie switched to the second die. This one was even more amazing because it seated the actual bullets. He poured grains of gunpowder into a bowl. The grains were dark, like tiny gray sticks of firewood, and made a soft purring sound as he poured. With a spoon, he gently sifted gunpowder into the cup on the scale. Slowly, the cup lowered; the other end of the scale rose. Howie eyed the balance marks, carefully spooning grains out of the cup, then putting a few back. Kole and I waited, wide-eyed, breathless.

"What do you think?" Howie murmured softly, as if even his voice might shift the scale.

"Now!" we said. "The lines are lined up!"

He grinned and nodded. He placed a tiny green funnel over the top of the cartridge and poured every grain of powder down into the primed cartridge. We weren't allowed to touch the shell now. That was fine, though. We had better jobs: I gripped a small heavy red box with a folding lid. It was two inches by four inches, thin cardboard, with metal-reinforced corners like a tiny pirate's chest. A drawing on the top showed two bullets that formed an *H* for Hornady Bullets. Inside were beautiful bright coppery new bullets, smooth and heavy and glowing.

Howie placed the powder-filled brass cartridge in the holder, nodded, and I handed him a bullet. Kole lifted the handle. When the handle lowered, out came our first reloaded cartridge. We passed it to each other, marveling at the magic we had performed. Reloading brought rifle cartridges back to life, giving them the power to acquire food and protect our family. Guns—and bullets—separated us from the very recent "old days" when people who lived here on this same ridge had to hunt with wooden spears, bows and arrows, and whatever else they could make by hand.

The wind moaned, sifting drifting snow over the moss and poles of our buried roof, hissing against the hot stovepipe. We glanced up from the reloading bench, for a moment pausing to listen as a fierce gust roared down from the north. In those moments our lives had two sets of walls—the walls of our sod igloo, and the walls made of sky, tundra, and mountains. Outside, the land seemed as big and dark and cold and barren as a lonely planet, with humans the scarcest creature out there in the night, and every other animal ranked in importance by what it might provide for our survival. The land those creatures moved across was wild and mysterious, and tonight turning colder and whiter with the coming of winter.

In the morning, with fresh ammo in his pockets, out onto it my dad again would go, small on a big land, leaving only a trail of tracks behind his dog team, a speck disappearing into the folds of the tundra as he headed out to gather more food and furs and firewood for his family. It was a life focused on gathering, and somewhere to the north was the cradle of that life. Caribou were the most important creatures, and they came from there.

CHAPTER 2

MINNIE GRAY

In the years I was growing up along the Kobuk, and later living near it, Minnie Gray was a steadfast source of calmness and kindness, compassion and wisdom—to the Ambler village, the surrounding region and beyond, and probably to every stranger she ever encountered. She was an elder before she was old. She told ancient stories, taught people how to take care of fish, meat, and furs; how to sew skins and collect birch bark, spruce roots, and medicinal plants; and so much more. Always full of graciousness and generosity, she shared her wisdom freely, without conditions. Minnie passed away in May 2019 at the age of ninety-four. It was a tough time for many of us, for a lot of reasons, and her presence has been greatly missed along this river.

This account was recorded in Iñupiaq in December 2014 and translated into English by Minnie's daughter, Helen Roberts, for the Western Arctic Caribou Herd Working Group in Anchorage. It was later published in the summer of 2015 in *Caribou Trails*, a newsletter of the Western Arctic Caribou Herd Working Group.

MY NAME IS ALIITCHAK. I GREW UP AND LIVED IN SHUNGNAK AND LATER moved to Ambler. When we were growing up, there were hardly any

Minnie Gray collects birch bark for making baskets. (Photo by Nick Jans)

The hooves of caribou, muskox, and moose are suited to their habitat and change with the seasons.

caribou in our area. We had meat like porcupine, ducks, geese, once in awhile caribou arrived but not often did caribou come around. Before caribou came around the men would go north with dogs to pack meat back home from the north. When the caribou started arriving, Iñupiaq culture and our elders told us not to block the caribou that are about to cross or coming in because you will lessen the migration. Don't interrupt the migration, this is important. After the first group crosses the river, then it is okay to hunt. When the caribou finally arrived, nobody wasted anything. The fur was saved and dried for winter use. In summertime they dried everything—no waste for winter use. When they bring the fur and legs, they use them for mukluks. Summer meat is dried along with fat. At Christmastime, *akutuq* (Eskimo ice cream) was made, and dried meat was used year around when caribou was brought from the north.

When I was growing up, caribou were not abundant, so using everything was very important. When we were young we camped and it was not easy, we had hard time hunting so we didn't waste anything. The fur was never thrown in the country. Everything was taken home. Our elders told us not to waste. Not to throw anything out. When you get caribou in the fall, then you dry it for winter. When it is spring you take the bones and smash and boil them. The fat on the surface is cooled and you dry caribou stomach inside out and then clean it and dry it. Then you put the bone fat in there and eat it with dried meat and dried fish. Yoi . . . good meat! The caribou fur is used in many ways. It is used for sleeping bags, mattresses—winter fur is better because it doesn't shed as much as spring fur. It is better for waterproof mukluks. That is how we lived way back in my days. Our elders always advised us not be wasteful.

Today there seems to be little respect from our young ones, and our children are not respecting and learning what they should. You take everything home and you dry it. Now days it seems like the young ones kill it and leave it out. We were told by our elders that is wasteful and that is not a thing to do.

When my cousin Sarah and I were camping, the caribou came around. I was 73 and I shot three caribou. We took them to camp, cut and dried. That is how I hunted at 73 and we dried the skins. Now today I am not able to do what I did at 73. I thank my Lord for providing all these years and I am happy to be here today.

CHAPTER 3

GOOD MEN AND BAD

Caribou had been on this land for a good part of forever, roaming the northern half of North America for two million years—long before any humans. Plenty of other professional predators were around, culling the herds and keeping them fleet-footed: dire wolves, saber-toothed tigers, huge bears, and a plethora of other fanged and hungry creatures that lurked on the landscape.

Only relatively recently—about ten or twenty thousand years ago—did the first two-legged hunters show up, crossing from Siberia. Those first people to arrive here took advantage of the ice age locking up water and lowering ocean levels, forming a land bridge across the shallow Bering Sea. These hunter-gatherers carried something the creatures of this continent had never dealt with before: weapons. Not rifles—not yet anyway—but spears and arrows tipped with sharpened ivory, bones, and stones. It had to be disconcerting, even confounding, I assume, for caribou and other animals to suddenly experience sharp sticks with hard deadly points sailing through the sky toward them. What could they be? What could cause such an aberration? On the

A grizzly sow and cubs displace wolves from their kill.

other hand, upheaval was already in the air; an ice age was ending, and like today global warming was on its way.

Around that time, mammoth, for one reason or another, went extinct. Evidence points to those sharp pointy sticks as contributing to their demise. They weren't the only Pleistocene animals to fade into the landfill of time. Saber-toothed tigers vanished, and the huge camels and the steppe bison and those tiny horses whose teeth and wrist bones we find along melting permafrost cliffs. The giant beaver and dire wolves of the era died out, succeeded by smaller relatives. Musk oxen stayed unchanged, patient, doing what they do best: hunkering down, waiting. And the caribou? On a landscape too harsh, cold, and barren for most other prey animals to even venture upon, caribou flourished.

After arriving here, humans—an invasive species brand new to North America— established permanent seasonal residences at the good spots, choke points for harvesting natural congregations of fish and birds, land and sea mammals, and other resources. These areas were often along rivers and coastal promontories. One of those was Paatitaaq (Onion Portage) on the Kobuk River.

Because of an alignment of low mountain passes, high tundra ridges, and rivers—the Hunt, Nuna, Cutler, and other river valleys to the north—and the elevation of the land to the south, and the proximity between winter and summer caribou habitat, Onion Portage was and remains one of the premier caribou crossings in the world. In the fall, huge herds often stream across the tundra, following veins in the land, ancient trails that lead out through the willows to the shore, where they pour into the water, lines of animals stretching from bank to bank, spanning the quarter-mile-wide river.

For a hungry people freshly arrived on a new continent, setting up camp for a long stay, with only primitive hunting methods, it is hard to imagine a more perfect place to stumble upon. When you climb the ridge at Onion Portage, you have a great lookout over the river and the surrounding tundra. Along the shore, the mixed spruce and birch knolls are welcoming, the walking open and pleasant. Underfoot, cranberries and blueberries grow plentifully, and of course wild chives, which gave this oxbow in the river its name. Firewood is available, and the land is relatively protected from the east wind by trees and shadowed from the howling north wind by mountains just six or eight miles away—mountains that provided the original occupants with jade, the only jade in Alaska, accessible on the surface, perfect for weapons, tools, and trade.

Down along the shore, the beach is knobby with big and small rounded rocks. It is these rocks that have kept the shoreline static, protecting the ridge from river erosion for the last hundred centuries and more. Archaeological evidence has shown that use of Onion Portage by Iñupiat Eskimos, primarily because of the seasonal abundance of caribou, began nine thousand years ago or so. (Incidentally, the same excavations showed evidence of periods when the site was abandoned, presumably because caribou, being caribou and unpredictable, had failed to return.) Seasonal hunting by local people continues today, although no one actually resides at that location anymore.

Before "contact," that odd extraterrestrial term for the arrival of Europeans, before metal and matches and all these modern material goods, Natives caught caribou here in snares and with spears, and by shooting arrows from kayaks into the swimming animals. Traditional hunting methods also included drive fences constructed of stone cairns, logs, and antlers; women and children hid behind rocks and undulations in the tundra, jumping out to scare the animals into the funnel of the trap. Hunters waited at the throat of the trap, often at the bottom of a hill, where caribou would pile up before a leader braved the water.

The modern romanticized portrayal of early nomadic Native life wasn't exactly the reality at Onion Portage, and surely other locations too. Instead, the Iñupiat there practiced what they were talented at: *waiting*. Especially waiting on animals. There in the path of the migration they hunted small game, fished, gathered from the land, and shaped jade and other materials into tools—meanwhile waiting, wisely taking advantage of the true nomads of the north, the caribou.

A handful of miles west of Onion Portage, along a straight stretch in the Kobuk River, another similar but longer ridge flanks the north bank, an outcropping of the same high tundra that funnels caribou on their fall migration and guides them back north in the spring. This three-mile-long bluff is called Paungaqtaugruk. It is here that my family settled in the fall of 1964, and I was born that first winter.

Kole and I race on homemade snowshoes. (Photo by Sasha Wik)

AS KIDS, KOLE AND I GREW UP SCOUTING THE SANDBARS AND CUTBANKS AND GAME trails, walking barefoot in late spring and summer and early fall, pressing our small insubstantial human tracks into sand and snow and soft mud, our eyes busy searching the horizons for animals and the sky for birds, and searching the ground, too— for black chips of flint, the broken-glass glint of obsidian, dark jade, and the gray of old antler and bleached bone. We hated shoes and loved going barefoot, searching for berries and birch bark and any sort of interesting twisted root or flat skipping stones. We picked up squirrel skulls to examine the rodent teeth, and rabbit turds, and stray rocks, and often we carried these treasures awhile, to later lose or leave, or stash somewhere. We had empty pockets and days full of time; why not carry a pretty rock for a few miles? In the spring when the snow was melting, we found dry moss to stop on and stand awhile to warm our toes. When we found a slab of slate, we built a campfire on it, or lay it over hot coals if we had meat to fry on the flat surface. Any exposed quartz we spotted, we scrutinized closely. Tommy Douglas in Ambler had told us quartz could contain gold in the cracks, like fillings in teeth, and we wanted to know if that could be true.

The animal trail we followed most often led through the trees, climbed the ridge past our sod igloo, and ended at the open tundra. From there the Jade Mountains towered in the northeast, close and friendly and familiar, and farther north the countless peaks of the Baird Mountains were like dusty blue diamonds. To the south, across the tundra, the Waring Mountains paralleled the river, and along their base a thin line marked the Great Kobuk Sand Dunes, gray in summer, white in winter.

Along that trail, near the crown of the hill, Howie had trimmed branches on a tall spruce. We called it his Lookout Tree. When he climbed up toward the top, we boys followed, climbing carefully below his boots, waiting for him to go higher and peering out through the green needles and nearby birch branches, catching glimpses of river, mountains, tundra, and the skyline ridge of Onion Portage.

Onion Portage was often in conversations, by then renowned in both the white culture and the Native one; locally it was known as a good spot for animals, and a likely spot to run into people, too. My family rested at Onion Portage along the winter trail when traveling to the village of Ambler. In the first years, travel by dog team was tough going with a heavy sled in drifted conditions, and Howie would often snowshoe ahead, and would pitch a tent at Onion Portage to allow us to warm up and the dogs to recover overnight. Later, after the early snowmobiles arrived and my family got one, we stopped there to have a snack, and to run back and forth to warm our feet and hands. In summer, traveling the river, we anchored our spruce plank boat along the shore, again to run around and warm up, and to pick the chives that grow there, and briefly to search the bank scoured by breakup ice for stone chips and pieces of arrowheads and antler sled runners, pottery, and broken jade scrapers, detritus lost or tossed aside by archaeologists and Stone Age toolmakers alike. We often found chips, or the most common finds—the bones of caribou, old and new and only a level above sticks, faintly disappointing to spot.

Four miles farther upstream, the long, high curved bluff of Ikpigruk erodes into the Kobuk River. The exposed glacial loess of those permafrost cliffs is steadily being eaten away during each season of water. Thawing, big and small chunks of soil plink and tumble and thunder down into the river to disappear into the dark depths. Passing in our narrow little boat, we would be in shadow, shivering in pockets of chilled air, inhaling the damp wet primordial odor and scanning the near-vertical cliffs, dreaming of spotting the giant bones and tusks of mastodons and mammoths. Something about finding pieces of prehistoric animals and the castoffs of ancient hunters was enthralling. It made those lives feel recent, as if the whisper of their passing was still on the wind.

Around us the land felt huge and endless, and the big walls of history felt close. One of those walls that fascinated Kole and me most was the divide between guns and what came before: bows and arrows and spears, slings and clubs and home-made snares and traps. I think most kids, even in cities in these ultramodern times, have a fascination with bows and arrows, spears and swords and such. With experience living off the land the way my family did, it was very easy—and at the same time daunting—to imagine a whole world without guns.

NIGHTS, UNDER OUR NEW KEROSENE LAMP, AN ALADDIN, I WAS HOMESCHOOLED AND learned in seventh-grade Alaska history about the next group of people to cross the Bering Strait, in the late 1700s. They were the Russians, and they famously brought copper-clad churches, cruelty, and change. My textbook was bound simply with white paper; it had black-and-white drawings and illustrations, and words, all of which painted the Russians as much worse than the Americans, describing hard men, evil occupiers who carried carnage up and down the coastlines of Southeast and Southwest Alaska, killing thousands of Aleuts and other Natives and decimating populations of fur seals, sea otters, and other furbearers. No connection that I recall was made between whales and caribou, and the only connection I knew of between the two was in my own life—how much I liked to eat caribou *paniqtuq* dipped in whale oil.

The Russians weren't known for venturing this far north and inland, and during that time period Iñupiat Eskimos, in family groups and small settlements such as those along the Kobuk River and other widely scattered points in northern Alaska, continued their traditional ways—hunting and gathering; residing in caribou-skin tents, log structures, and small subterranean sod dwellings; and using spears and arrows and other time-proven methods to catch and kill caribou. Trade between various groups, hammered out over millennia, was an essential part of survival, and across the north, trade routes covered great distances over rough-traveling country. Tribes fought and raided each other and conducted wars, and trading parties often met at fiercely protected territorial boundaries, and—if I know anything about the Native grapevine, which was uncanny and near-magic even before telegraphs, CB radios, and Facebook—I assume a few fragments of metal and a tale or two of those pale marauders *must* have found their way to the isolated inhabitants of the interior and Arctic coasts. Regardless, when it came to caribou and hunting and human reliance on caribou, historical accounts suggest that here in the north, little had changed by the early 1800s.

It's surreal to sit here at my window now, to glance up from my glowing screen and stare out over this big river and realize that all that history was a mere two hundred years ago.

BRITISH AND OTHER EUROPEAN EXPLORERS FOLLOWED THE RUSSIANS NORTH, MANY OF them searching for the Northwest Passage, a fabled ice-free route connecting the Atlantic and Pacific Oceans. Now, with global warming, luxury tour boats and even jet-skis cruise high-latitude waters where those brave explorers struggled in unfathomable pack ice, where many died, some famously—not for their deeds as they had dreamed but instead famous for the cold, dark, and lonesome deaths they met in the polar north. Today, in the brilliance and comfort of hindsight, it is again surreal—sad and unnerving, too—to realize how those men weren't so wrong after all. They were only off by these two hundred years.

The explorers who did return home carried information, including one surprising tidbit of news. There were whales in the Arctic Ocean. It was a small detail—except whales brought commercial whalers, and this newest batch of newcomers ultimately brought guns.

By 1850 whaling ships were just west of my birthplace here, offshore and sailing up the Bering Strait. Onboard were the newest Outsiders, from mysterious places such as Sag Harbor, New Bedford, Mystic, and other faraway ports; Americans and other rugged men lured from coastal communities around the globe to labor for the big oil companies—big oil in those days came from living creatures, mostly sperm whales. (Coincidentally those nineteenth-century seekers of oil came halfway around the world to the same Alaskan coastline that ConocoPhillips, British Petroleum [BP], and other modern oil companies have descended upon.)

Sailors signed on for journeys of unknown duration—which often turned out to be two years—setting out in autumn, sailing around the tip of South America, hunting whales in the South Pacific until spring, off-loading their winter catch of oil and baleen in the Hawaiian Islands, and then heading north for the summer season in the Bering and Chukchi Seas. In the Arctic, as fall came on, a year into their voyage, the men were damp with salt water and whale oil; jammed in reeking, creaking, cramped quarters; fed salted horse and hardtack and whale meat. Cold and sick and homesick, they battled storms and ice, unknown shorelines and constant uncertainty, surviving—and often not surviving—and no doubt carrying that basic array of items men have carried on similarly dangerous forays down through the ages: weapons, disease, alcohol, and of course sperm.

Good men and bad, I assume—white, black, Portuguese, Hawaiian, and other people garnered from across the oceans—these were no handpicked ambassadors of the lands from which they came but a rough unsorted crowd who by a hiccup in history happened to be the ones to make initial widespread "contact" with the Natives in northern Alaska. Natives who were still living an ancient aboriginal way of life in a harsh cold homeland where over hundreds of generations they had grown to depend on caribou more than any other creature. Grown to depend on the meat and fat as sustenance, but much more importantly on the insulating fur that caribou unwillingly provide.

AS SPERM WHALES BECAME OVERHUNTED, WHALING CREWS TRAVELED FARTHER NORTH. They began harvesting other species of whales, including bowhead, which the

After swimming the Kobuk River, caribou navigate eroding permafrost bluffs at Ikpigruk.

local Natives hunted in the Arctic Ocean from their *umiat* (small skin boats) and relied on for food and for oil for their stone lamps to cast faint light and heat in their tiny sod igloos in winter. Whalers also turned to walrus for oil and ivory, and killed thousands along Alaska's coast off Wales and Point Hope and other settlements.

Sailors were sent ashore to barter and hire replacement crews. Suffering from lack of blubber and meat from whales and walrus, and also greatly desiring the new trade goods, Natives began hiring on to the ships: the men to hunt caribou and the women to make clothing. Whalers preferred caribou meat over that of marine mammals and had discovered that Native clothing fashioned from caribou skins was far superior to their wool and oilskins, especially for winter conditions.

Trading posts were established. The Outsiders began staying longer, later in the season, and soon ships were trapped in the ice, stranding distraught men who were forced to overwinter and rely on the indigenous population to feed and clothe them. Pressure on caribou increased from Natives and whalers hunting for themselves, and from Natives hunting for the whalers. Meanwhile, trade items from the Outside—tobacco, liquor, sugar, flour, cloth, needles, metal, and other materials—disrupted traditional trade networks, and the introduction of alcohol further clipped connections, adding disillusionment to distress as newly arrived diseases spread rapidly, decimating settlements and isolated groups, compounding the rapid deterioration of the Native lifestyle.

In a shockingly short time—just a handful of years, a few decades, a generation or two—ancient traditions and values were in tatters, shredded by a blizzard of new technology.

THE WHALERS INITIALLY WERE RETICENT TO TRADE FIREARMS. BUT THEY DESIRED FURS AND meat, and by 1870 trade had put rifled muskets into the hands of nearly every Native hunter across the Arctic. With the introduction of rifled muskets (much more accurate than the previous smooth-bore muskets)—and soon after, breech-loading repeating rifles—customary hunting practices were largely abandoned. Harvesting animals remained of utmost importance to Native peoples even as methods altered with new, more efficient means. With a history of hunting in one of the harshest environments on earth, using only primitive implements, the Eskimos had honed astounding hunting prowess. Now those skilled hands held rifles, and those weapons were pointed at caribou. On the tundra, an era suddenly ended and a new one arrived. Overnight, the Stone Age became the Rifle Age.

What happened next in caribou country was, to a certain extent, buried by the snowstorms of time. Native Yup'ik and Iñupiaq cultures had no written language. Explorers, miners, and other Outsiders were few and far between, and were exactly that—Outsiders—with limited or no baseline information to compare their brief and passing observations to. And Alaska was immense beyond comprehension, populated by wolves and bears, moose and caribou, and other creatures, all busy carrying on their natural interactions largely out of sight of the tiny specks of human habitation.

Historical records that do remain tell indisputably of a devastating crash of the northern caribou herds in the second half of the 1800s. Native oral accounts tell of caribou being absent or scarce. Journals scratched down by miners and explorers speak of very rare sightings of "deer." Those records don't say why the herds crashed. The chips of information left behind prove that those seemingly improbable stories told to Kole and me as boys were true—and show how lucky we were to be gnawing on a steady supply of sweet, fat caribou bones, to be sleeping on soft caribou hides, and to be wearing warm mukluks and parkas. And to be able to watch groups of animals flow across the tundra, from the north every fall, and from the south every spring, like a clockwork of the land.

Strange as it seems viewed from this present of plenty, there was a time, very recently, when caribou were rare and even nonexistent at Paungaqtaugruk, Onion Portage, and across large areas of western and northern Alaska. Minnie Gray and other elders were right: they knew their recent history; they had lived it. The elders had watched as the herds, trusted like the seasons to return each year, had faded rapidly into memory. For half a century or more, this area had been nearly void of caribou. How lonesome spring and fall must have become, how hungry and empty and cold everyone's hearts. How terrible it had to be for the men daily searching the horizons, walking the tundra, staring across the vast distances, waiting, trying again tomorrow, trying ever harder to spot the distant dots that had to be there, had always been there, but now inexplicably had vanished. And the women, hushed, home awaiting word of a successful hunt and the instant joy and rush that word of fresh meat brings, and with it the fat soup, meat to dry, and skins to tan and sew into warm clothing for their families. And the children too, waiting, quiet, watching their parents' faces and seeing the defeat carved deeper each day.

CHAPTER 4

BEING LUCKY

In Kobuk village, in August 2016 during a rainy spell, Nina Harvey and her sister, Mildred Black, invited me and my friend Diana in for boiled moose bones. Nina and Mildred are Iñupiaq elders, both in their eighties, and they haven't exactly lived lives of leisure.

Nina's is a modern government-built house, with a kitchen bright and cheerful with fluorescent lights, a fridge, and a sink with hot and cold running water. They shouted approval when we came through the door. The sisters were sitting across the table from each other in straight-backed chairs, old ladies, small and hunched low behind their plates, smiling over huge steaming bones.

Nina gestured toward the stove, and they were pleased when we quickly took plates and moved toward the stove to peer into the soup pot. Right away I spotted sawed leg bones, with big circles of fat *patiq* (marrow). My mouth started watering and I forked a knee joint out, being careful the big bone didn't slide off my plate. I reached for my sheath knife and glanced around for salt.

Mildred is a good storyteller. As she cut at meat and fat, chewing and talking, she told about her and Nina, as young girls growing up at the mouth of the Mauneluk River. Diana and I chewed and listened. Nina mentioned the recent rain, which had interrupted berry picking

Female humpback whitefish hanging to dry, with stomach and eggs intact to ferment (Photo by Pearl Gomez)

Moose have expanded their range farther north for a century or more as the treeline advances.

and caused the ladies to cease seining for whitefish. We all lamented how tall the willows and alders are growing nowadays, so thick in places that we can't spot animals and can hardly recognize known spots along the shores of the river.

Mildred is tiny and incredibly stooped. Nina is wiry and agile, constantly moving. Their conversation meandered back and forth: to the fact that no caribou had come yet this fall; wolves were plentiful; young people don't fish enough anymore—they watch too much TV, and drink, and do drugs, and don't know how to survive. "They gonna starve when hard times come again. They need to know how to fish."

I listened to the concerned old women, noting that Mildred used the word "when" and not "if" talking about hungry times coming. I cut at my moose bone, wishing now that I'd chosen one with a little meat on it, not just all fat and cartilage. I enjoy Mildred's stories and gently steered her back upriver, and back in time, asking: "When you were kids, when did you first see a moose?"

She told a tale of no moose in the country—back in the forties—and how one day her uncle got a strange dark animal, a moose. One fun part about Mildred's stories is that nearly everything that took place more than fifteen or twenty years ago, she refers to as "back in the forties." Once you accept that as a catchall phrase, not an actual date, you can't help smiling each time she says it, and marveling at the cool way that so many seasons have settled in her mind like fallen leaves.

Her statement about moose and their uncle fit with what I was pondering. Basically, had moose been here since time immemorial, or had they arrived relatively recently as they had around Kotzebue? To me their recent arrival suggested that vegetation—willows and other shrubs—had been increasing steadily here, even, ahem, back in the forties.

From there my thoughts scattered, wading through what we've been experiencing here in Northwest Alaska: much warmer falls and winters—so warm that we have rain blizzards in winter now—and changes in wind, water and ice, and weather patterns; changes especially in the rapid overgrowth of vegetation; changes in the movement of animals. Drastic changes in us humans, too—unimaginable changes, if viewed from the recent past—and yet we still hold on to uncertain connections to the land, hunting and fishing and gathering.

It was late August, and not surprisingly my mind moved to caribou, arguably still the most important animal to most local people, regardless of the inroads made by boxes of technology flooding north—the Xboxes and the Amazon boxes, the manufactured homes and steel shipping containers, the boxes of Pampers and cases of Pepsi, TVs and telephones, and all the rest. For a minute I chewed and cut at my big bone. In my cluttered head, I was sifting through buried accounts I'd read of the early 1900s along the Kobuk: the starvation and suffering, the diseases, and the dearth of caribou in the years following the crash of the herds in the late 1800s.

"What about caribou?" I finally asked Nina and Mildred. "When did you first see those?"

I'm shy. Especially around elders. Over the years I've mostly kept my thousand questions to myself. Like many members of the "younger generations" here, I'm deeply embarrassed by my lack of knowledge, by the huge and obvious gaping holes in what I know about the land and animals and the Iñupiaq place names and proper seasons for gathering this or that (for example, which species of whitefish to preserve when and in which way), and lately all of that is made worse by the fact that I'm white, and why am I asking? What am I planning to do with their traditional knowledge? My eyes stayed down, on my knife, as I carved at another bite of tasty fat. The moose was fresh and so good, a young bull. "Were there caribou around when you were kids?" I murmured.

At times Nina can't hear too well, but Mildred can hear and think and remember just fine. Instead of answering directly, she did that beautiful thing that elders often do—she began to tell a story. In this case the story was about her as a seven-year-old, back in the forties, when as girls they first saw a caribou. The men had all gone hunting, searching for small game or whatever they could find. They left the women to tend camp. A lone animal with antlers showed up on the sandbar below their camp. The women had a .22 rifle and one bullet. An older sister took the gun and the bullet, and she moved quietly downriver through the brush. After a long while, the girls and women in camp heard a tiny distant report from a rifle. They rushed to grab their *ulut* (curved Iñupiaq knives).

They hurried along the shore. As they came out of the brush at the upper end of a sandbar, they could see their sister. Lying near her on the sand was an animal they'd never seen before. It was large and very dark, with a rounded stomach and chest and large black furry antlers. The hooves were black and the legs and flanks dark with short summer hair. The women and girls had never worked on a caribou, but of course they cut it up and hauled the meat back to camp.

Sitting at their table, I wondered if that animal had been a lone stray, or had that marked the beginning of caribou returning in larger numbers to this country? I kept my question to myself. In this region, asking too many questions is often considered bothersome, and definitely white. I couldn't help smiling, though, and asking, "What did the men say? When they got home."

Mildred and Nina glanced up. The two old ladies grinned and shared a glance over their moose bones. "Lucky!" Mildred said. They laughed happily, and then we all laughed again, in part because of how one word so perfectly sums up the complexities, effort, and uncertainty involved in finding food on a hungry land.

The telephone rang shrilly. Nina leaped up to answer it. "Mail plane coming!" she shouted. "Ravn fifteen minutes out!"

The old days faded from the room. Now, suddenly back in the present, with a plane to catch, I glanced at my backpack and boots by the door. I wiped the fat off my knife onto my finger and licked my finger, sheathed my knife, and quickly checked the time on my iPhone. Still, I could see that bull caribou, dark on the sandbar. I could see the knobby knees and black hooves, the soft black velvet antlers, and the animal's dark thin summer coat, and on the sand beside the caribou a bright pool of coagulated blood, with flies buzzing and gnats and mosquitoes biting, frenzied over the smell of so much fresh blood, and the soft green summer leaves on the willows, fluttering in a warm breeze, and the quiet current of the river flowing past. And I could see how lucky I've been to have lived here, all the seasons with caribou flowing through my life.

And for a few more moments, I sat at Nina's table, comforted to remain lost in the years, to picture the continuous stream of food, the thousands and thousands of meals these two sisters, and I too, have harvested from this land.

CHAPTER 5

HUNTING ON THE FALL TUNDRA

In the bright fall morning as I come out of my sod house and pass the protection of the *qanisaq* (entryway), the wind greets me with a sudden sideways shove. The gusts are cold, stunning, and seemingly alive and angry in their intensity. I'm barefoot and shirtless, and my eyes are already hunting—searching the tundra and riverbank for animals and checking the weather—letting the sky steer the course of my day, as I do every morning.

The brilliant blue and the shards of sunlight rippling through the last tortured yellow leaves are momentarily blinding. Instinctively, I glance around for a bear, and stumble off the gray boards of the entryway to find a tree to pee behind.

The spruce trees and birch branches bend and whip, fighting the roaring wind. I stare up and around at my childhood companions, these trees born here on this hill like I was; we are friends in a way. We've survived many of the same storms in the last half century, and somehow survived that biggest blizzard of all: life, with its gusts of luck and loss, and lately torrents of change. Here, around me, it appears that our overheated summers, melting permafrost, and new warm winters

Bulls work their way through dwarf birches and spruce on the tundra.

have been kind to these trees, and I marvel how for now they seem to be winning the war with the wind.

Directly below, down the steep ridge, the quarter-mile-wide Kobuk River flows past, dark blue and choppy today, with the wind sweeping cat's-paws over the surface. Far across the water, above the shore, the black speck of a raven soars east, high on the wind. I watch to make sure he signals no messages. We both want me to kill a caribou, but it is growing harder each year to spot them in the jungle of new tall brush bearding the riverbanks. And the herd doesn't seem to be migrating through here in September anymore.

My gaze searches the shores and sandbars for any dark dot, any white speck, any movement. I remind myself of Howie in those moments, and another important hunter in my life, a man named Clarence Wood. And I remind myself, too, that much of my family's diet when I was a kid was exactly that: *anything that moved.* That intense subsistence mindset endures as part of my life, and I hurry back up onto the gray boards, stopping to peer at my parents' old thermometer. The red line is at 33. The sky has me hoping to hunt today, for meat and for more cranberries, but I'm addicted to all my favorite foods that each season offers, and no frost means the blueberries will be on the bushes for another day. Also, this cold wind might help bring caribou from the north.

I'm shivering but I can't help grabbing my binoculars hanging from a nail in the *qanisaq.* Quickly, I glass to the north. Across the miles, the tundra glows in fall splendor, burgundy with splashes of orange and spears of silky green spruce and the distant golden dots of lone birches. I fight the urge to run for my Nikon camera; instead I stand and scan the flats behind my family's old sod igloo. Nothing is visible, although nowadays I could miss a herd of a hundred caribou in the thickets of alders and giant "dwarf" birch that have overgrown the flats. Farther out, toward the mountains, I see what might be tiny grains of rice or caribou on the open tundra—but the wind shudders my binoculars, and my eyes water, and I can't be certain.

Inside, I crack the stove draft and stand close. The black barrel makes ticking sounds as the metal warms. My teenaged daughter, China, stirs on her bunk. I watch to see if her tousled curls appear from under her down sleeping bag. I decide

Birch leaves turn in autumn as the tundra's fall colors begin to fade.

suddenly not to wait for a kettle of water to heat, to skip coffee—a rare occurrence— and instead I slip into hiking boots, a sweater, and a windbreaker. All the hungers I feel are to hunt. At what used to be my dad's workbench, I pocket reloaded .270 cartridges. The simplicity of the old brass reassures me, and helps tie the past to this uncertain and changing time, the present. One shell slips out of my fingers, rapping on the beaten boards. I glance at China's bed, and over at my wife, Stacey. Their breathing remains steady; stealthily I grab my camera, pull on a hat and gloves, shoulder Howie's old rifle.

The game trail that Kole and I as kids used to sprint along barefoot still leads through the trees to the top of the ridge. When I come out onto the open tundra, the sun dances on a million red and orange leaves shaped like tiny seashells, an ocean of dwarf birch leaves glowing in fall colors. I pull the gun off my back, cradle it in

Blueberries dot the tundra.

the crook of my arm, and move quietly through the wiry shrubs. Once, I walked head-on into a grizzly here; oftentimes I've met a moose on this path, and of course countless caribou over the decades.

Ahead, the tundra stretches north to the Jade Mountains. They tower dark blue and burly against the sky. I stop and glass for caribou. There are none. Nothing moves except the wind in the trees along the edge of the ridge. The land is the way I know it best: huge and rolling to faraway foothills, rising to mountains and beyond to more and more mountains. There is no hurry now. I know how empty this land can be, and how suddenly animals can appear. Hunting, for me, means being out here. I pause to pick a few crowberries and a handful of blueberries. The leaves have fallen and the berries hang in unobstructed view, beautiful tiny balls of chalky blue. The taste is perfect—sharp and sweet and intense as this day.

I stride northeast, moving fast on caribou trails that fan and fade across the uneven grassy tussocks. My chest fills, inhaling the aroma of the tundra. I've walked these game trails a thousand times, and swallowed every kind of edible berry that grows on this ground, hunted every animal, suffered for photographs, and set off on journeys from here that nearly took my life—as it should be in the wild. I feel my lifelong home all around me, harsh and indifferent, yet supportive, and my heart aches and flows over with love for this land.

My eyes know intuitively to no longer scan for dark dots as they do in late August, but to instead search for the white manes of mid-September caribou. Something catches my gaze, a shape or color out of place, far away on the tundra. Quickly, I raise my binoculars. Tucked down in the grass and partially hidden behind a rise, a handful of caribou is bedded down. At first only faint white specks show—a few faces, necks, and flanks. My eyes strain to take apart distant details. Suddenly my muscles tense. For long moments I'm too focused to breathe. Behind the visible animals is a patch of willows, except it's not willows—it's a thicket of curved antlers. A group of big bulls is resting there. They are the ones I want, the best meat of the season.

The herd is much farther from home than I usually hunt on foot. But the sky is bright and in these modern times—like those old forgotten days—none of us can predict if more caribou will come this season or we'll be left hungering for them. We no longer trust what we once trusted most from the land—the caribou migration, and now even winter itself, with ice to travel on. I hunch my shoulders down out of sight, cradle my rifle, and set off for a draw to drop into to work my way east.

For an hour I'm intent on the sneak, moving as fast as I dare, crouching below hummocks and taking cover behind new clumps of alders and willows and stray young spruce. Finally, I'm belly-down, crawling over and between grassy heads, snatching a few hanging berries to swallow, wriggling forward, sweating, clutching my binoculars and gun. My elbows and knees are wet from tundra puddles and stained blue and red and purple from berries. I'm close now, and positioned downwind, but still all I can see are antlers. Cautiously, I rise to crouch behind a baby spruce tree. I aim, and wait. The north wind shakes my rifle.

A few heads are visible now, but no clear shot. What I really want to see are butts to scrutinize for fat. Fat still matters most. Skinny meat no longer means suffering and starvation, but even today it borders on taboo to bring home a skinny animal. Across the modern Northwest Arctic, the first two questions asked of a hunter have remained unchanged: You catch? Was it fat?

My sweat turns to chills in the wind. The sun has a bit of warmth, but I'm growing cold. Suddenly a bull is standing. I hadn't realized how close I'd gotten to the animals. Most of the herd is over the rise, resting still. At any second they might scramble up and be gone. Quickly I aim. I don't examine the caribou for fat, just squeeze the trigger.

The bull crumples and instantly I'm bounding across the uneven tussocks. Pulling my knife. Grasping the huge antlers. Stabbing in at the base of the skull to finish it off. I stare around, exalted, and thankful beyond words. Finally! Lucky! Fresh caribou! This is what so many people in our region are hungry for in this season: big bull caribou.

The group runs a hundred yards, slows, and stops. I wave my white cotton gloves over my head. A few bulls face me and then walk closer, curious. I feel that familiar temptation: to shoot another, and another, to get meat while I can. I make myself pause, and glance back the way I came, to remind me how far I am from the house. I bend down and begin my work. The bull is large and very fat. One is enough. And

China hauls caribou hindquarters home on a pack frame.

A gray jay goes straight for the fat, stashing food for winter.

besides, our Septembers are not the Septembers of my childhood; the bears are still out, and the flies and wasps and other insects too, and the weather is much too warm to store meat.

By the time I am done gutting the bull, the herd is a string of distant dots far to the east on the tundra. I cut alders to cover the carcass to keep ravens off the meat, shoulder Howie's rifle, and then, using the brisket as a tray to hold the liver and heart and *itchaurat* and tongue, I start across the tundra for home and pack frames and my family.

WHEN THE THREE OF US ARRIVE AT THE CARIBOU, STACEY PUTS DOWN HER PACK FRAME and looks back across the distance we've come. "Is this the farthest we've ever hauled one?" she asks.

I shrug. "Maybe so." I pull out my knife, point with the blade. "See how fat he is?"

China kneels beside the caribou and opens her knife. We're about to start skinning when Stacey suggests that she and China skin and butcher this caribou—with guidance from me, but no help allowed. Surprised, and a little chagrinned, I step back. The animal is already gutted, so I point and suggest first cutting off the lower legs at the wrists and ankles—on a caribou those are the joints a foot or more up from the hooves. I explain that there are three joints on the hind ankle and how the lowest one is best, although harder to find.

They struggle with the boney legs. "Geez. You always make everything look so easy," Stacey comments.

They finish cutting the front wrist joints. I gesture for China to split the skin up to the jaw and skin it back a few inches on each side to where the skin becomes less attached. "Now use your hands to skin." Instantly I picture Howie long ago teaching Kole and me to use our fists, to quickly and cleanly skin a good portion of each caribou. I demonstrate to China; I try not to do too much. They quickly shoo me away. I pace the tundra, incredibly restless and drawn to touch the animal, use my hands and strength, feel the warmth of its body under the skin, admire the beautiful fat, meat, and bones. Hunting for me has always been far more than pulling the trigger.

I eat blueberries, glass distant valleys, and contemplate whether all these years maybe I've done too much of the work when it comes to caribou. Maybe I've always had them do too little. But at the same time I've had such doubts about what skills of mine will matter to my daughter. Now she's taking classes in urban planning, of all things, at Stanford University. She wants to live in Berlin. Where, I wonder, will homemade snow ice cream, gutting geese, and skinning caribou fit into her future?

I find myself grinning. *How can they work so slowly?* Howie taught us to work fast!

Finally, I borrow Stacey's Samsung smartphone and attempt to take photos of blueberries and the fall colors, the tundra and sprawling mountains to the north. I keep pressing the wrong button, shutting it off instead of snapping a photo. A stray thought comes into my mind: *Could it be possible that some people might find gutting and skinning a caribou to be as confounding as I find phones and computers and electronic menus? But wait, how could that be? Skinning an animal is so elementary, so normal.*

"Did you bring a sharpening stone?" China shouts to me. She's nervous, concerned she might cut the wrong thing, mess something up, and her father will notice it all in a glance. They are swatting at no-see-ums, blinking against the bugs, and brushing their faces with their forearms. I stride over and touch up the edges on their blades using the back of my knife for a steel. My arms are splotched with red dots—bites I've ignored. I gesture for Stacey to roll the caribou to the side, and for China to lean low and reach in to skin all the way past the backbone. Now I have her roll it the other way, and Stacey finishes skinning on her side. I show China how to bend the tail to find a joint, cut it free, and then how to cut around the butthole—a circle, pressing hard against the bone on the inside of the pelvis, and then pulling out the pee sac and poop chute and flinging it aside for the ravens.

Now only the neck is still attached, which they relinquish to me to sever. The skinning is done. I show them how to find a disc on the inside of the backbone at the waist, and cut the animal in half, and then how to pry the shoulders out away from the rib cage and cut them free.

Ravens pass overhead, black wings panting against brilliant blue. The birds caw and comment, pretending patience. One swoops down and perches in the top of a nearby spruce. Its black eyes are watchful, as judging as mine in this whole process.

We lash the meat to our pack frames and onto an orange plastic kiddie sled I brought along. I say a last thank-you, take a last glance around the kill site for anything forgotten, and we set off trudging across big tippy tussocks. The day is bright and sunny, the tundra lit up in colors, and the sun offers warmth against the wind. Our loads are heavy, heavier every step, and we stagger on the uneven tundra. The sled catches and pulls horribly behind me, snagging and dragging and tipping over. Slowly, the kill site grows smaller in the distance. The gut pile is a gray pillow beside the upside-down head where the spikes of the antlers remain pressed into the soft tundra.

After a mile, we are on smoother ground, and around our feet small new spruce are everywhere, rapidly populating this old tundra in the increasingly warmer seasons. Peering closer to the ground, I see thousands of them—like crowberries, but no, these are tiny white spruce. In another decade this will be another new forest. Ahead, the young trees are thick and green and healthy where the tundra was bare and open when I was a kid. Quickly now we're shaded in the tall birches and towering spruce forest on the game trail leading down the ridge. Ahead, in a few hundred yards, the cache will come into view, and the outhouse and wood-pile and sauna will appear through the birches, and we will lower our loads and rub our shoulders and begin to hang the meat. Maybe it's noon now, maybe later. More work waits in front of us, fulfilling work that makes the day feel long and sustaining, satisfying and gratifying and good.

LATE IN THE EVENING, I HAUL IN A LAST ARMLOAD OF KINDLING, AND PUT AWAY THE BUCKET of cranberries we picked in the afternoon, and stash the quarters of the caribou in our log shed for the night. Inside at the table, China and Stacey spread out our feast: cranberry sauce, greens and potatoes and carrots from our garden, and fresh meat. We're hungry and gather around the Dutch oven for boiled brisket and tongue, our traditional first dinner after getting a caribou. We fork fat short bones and slices of tongue out of the broth. Our fingers are greasy, our lips, too, and the handles of our knives. It's hard to stop eating. The flavor of the meat and bones and broth is exactly as it should be, rich and fat and full with the taste of life, and with it the memories of a lifetime of meals gathered here. There's something bigger too, a connection through food that caribou provide, an indelible intertwining of our separate species here on this land. And there's some more powerful force, too, that caribou bring to our lives. It feels like love, friendship, or a nameless kind of companionship.

CHAPTER 6

CARIBOU SOUP

If you are a hunter, or happen to know one who has handed over a roast or some soup bones, be aware that caribou is not a forgiving meat, and many folks not familiar with it say caribou tastes gamey, dry, or like liver. The meat is not marbled with fat in the way that beef or pork is, or even goose, bear, or musk ox. Caribou are astonishing athletes; they run for a living, fleeing daily at unbelievable speed across tundra and snowdrifts, up mountain slopes and through miles of slush and brush, swamps and ravines. Compared to a caribou, a chicken from the supermarket is a helpless morose jailed jellyfish.

For that reason, it helps to have some hints about how to cook this wild meat. My family's traditional first taste of fresh caribou—while our hands are still crusted with blood from the kill and before we continue preparing and storing the carcass—is fried liver and heart. Slice the liver no more than half an inch thick; slice the heart a bit thinner because the dense fine-grained meat cooks slower. You'll want to cook it hot, in a little olive oil if you have it, or butter, or caribou fat—hopefully with sliced onions and fresh *itchaurat* if the caribou came with some. Fry the *itchaurat* and onions first. Then push them to the side and add the slices of heart. Sear one side, the second side. Now add the liver slices to the pan. Flip the strips over again, cutting into a piece

A brisket hangs from a poplar tree to dry and develop a crust so the meat will keep.

of each to see if they are nearly done. Bring the hot pan to the table. Start eating. Don't wait for stragglers. Eat fast. It's still cooking, and you've got lots of work to do when you're done eating.

After that meal, go back to working on the animal: skinning and hanging the meat in the shade—somewhere cool and inside screening or a fishnet, or in game bags if you have them, protected from gray jays and flies if possible. The brisket and tongue can be set aside for dinner tonight. Caribou brisket and tongue are the first places a caribou puts on visible fat—in addition to the *itchaurat* and a ring around the top of the heart—and these are choice pieces in local Native culture and are used for the quintessential meal of this region, caribou soup.

The brisket is long and V-shaped, made up of short ribs that you cut free at each joint from the flat center bones. After the small bones are removed, saw or score and break the flat sternum bones crossways into five or six sections. A thick flap of large-grain meat pads the sternum; cut this into squares or leave it attached to the sternum bones. Toss the bones and meat and tongue into a Dutch oven or large soup pot and add plenty of water. Season with salt and pepper, garlic, and maybe a couple splashes of Worcestershire sauce, or whatever you prefer. Bring it to a boil.

Simmer slow for a few hours or more. Add a little salt near the end if the broth needs it. The meat should be very tender and the broth tasty and fat. Keep this meal simple. You hunted today, and tonight you want to taste the caribou. Because this is soup, and because the brisket and tongue are fat, you can get away with the meat not being aged.

The rest of your meals are best divided into two categories and treated accordingly: soup meat, and steaks and roasts. Soup meat can be cooked fresh, or frozen fresh, or hung and aged. It will taste great as long as the meat is cooked with fat, especially fat in the form of bones. From an ancient Iñupiaq perspective, fat is all important. Everyday caribou soup—made with ribs, backbones, or even sinewy lower legs—is similar to the brisket cooked the day of the hunt: put a little oil in a big pot and sauté an onion; add diced meat and sawed-up bones along with plenty of boiling water. Don't leave out the bones and their fat and flavor. Even the toughest coiled meat of the lower legs, diced in small pieces, will turn out tender if your broth has bone fat.

Bulls feed and get along peacefully before the rut.

After you season and simmer for a few hours on low, then consider variations: some cooks toss in elbow noodles and canned tomatoes. I like to put in a handful of barley and another of wild rice that a Chippewa friend, Angela, sends me from the Lower 48. Brown rice or plain white rice work too. Don't put too much though—it's soup, not stew or goulash. Add chopped potatoes, carrots, and celery if you have them, even kale. Taste it again but be cautious. There will be fat on the surface; don't burn your tongue.

Steaks and roasts require the meat to hang and age for a few days minimum. Aged properly, the meat will be tender and flavorful, and you'll never go back to dry, gamey, liver-flavored cuts. As the meat ages, the steaks will just keep getting better, and each day you'll look forward to dinner more and more. If you fry meat the first night of the hunt—even the tenderloins—the meat will taste okay but will be chewy as rubber bands. (Speaking of tenderloins, remove them promptly after gutting the caribou. Otherwise blood will pool there and coagulate in this splendid fry meat. And don't let the name "tenderloin" fool you; this animal has just ended a lifetime of daily fleeing twenty-five to forty miles per hour over tussocks—like sprinting on bowling balls—and it only makes sense that it takes a few days for the meat to relax.)

In the best scenario, you hunt in September when bull caribou are big and fat and weigh hundreds of pounds. In September, the meat no longer has that murky flavor from their summer diet of willows and whatever else makes the meat taste a bit like mushrooms, and it doesn't yet have the hormone stink that begins with the rut in the first week of October.

If you hunt on the tundra, you should watch the animals before you choose one, to observe which are healthy. Look at rear ends, to see if they are round and fat. If you do hunt in the river, remember, the fattest animals ride low in the water, not high as you might expect. A neck shot is what you want, or behind the ear. A chest shot wastes meat—often the heart, ribs, and part of one or both shoulders. Chasing caribou causes adrenaline to flood into their veins, which can cause an off flavor. Take your time stalking and walking and waiting. We're talking about gathering food—something best separated from gathering manhood.

Once you've killed a caribou, gut it promptly and cleanly—no guts on the meat, and especially no sand. Be extra careful if you're along the river; you can trim meat, but you can't wash off sand. Haul the meat home in the largest pieces you can carry. The reason for big pieces is to keep as many of the muscles and tendons as possible unsevered; this is important in the aging process. Often, when packing meat, it's easiest to cut the caribou in half at the waist. If you choose to do so, at this point you must decide if you want to save the hide or cut it in half, too. Left unskinned, the meat stays cleaner, and waiting a few hours to remove the skin means the back fat will remain beautiful pearly white. Cutting the front quarters off the rib cage in the field makes the animal easier to carry and doesn't hurt the aging process. Skinning with your hands—where possible—is always best because it keeps the most fascia, meat, and fat on the animal and prepares the hide best for drying. Fascia on the meat also protects it and provides a handy layer to trim later if necessary to remove leaves, hair, or twigs.

Once hanging, the weight of the meat stretches the muscles against tendons still attached to bones. Rigor mortis sets in, clenching the meat and internally tearing the fibers, and eventually relaxing again in a few days. This is why you don't want to "bone out" the meat in the field. Also, now enzymes begin breaking down the tissue of the muscles, tenderizing the meat and adding flavor. A cool temperature and circulating air are both important, and in the right conditions you can leave meat hanging for two or three weeks. If you hunt in winter, you can mimic some of this process in subzero temperatures by digging a big hole in the snow, placing the skin in fur side down, and laying large sections of your caribou on top of the skin. Fold the hide closed, lay a tarp over it, and bury it under more snow. The meat will stay thawed and age for a few days or even a week.

By the third day, you'll notice your fry meat tastes better and is more tender. By the fourth day, you'll find yourself wanting to fry up a second pan. A bowl of fry meat in the fridge—if you have a fridge—can be used nightly for a week for steaks. As long as it doesn't go sour from being wet or too warm, the meat will keep tasting better and become more tender.

Cut your fry meat with the grain to keep in juices. Remember to cut some pieces thinner, some thicker, to allow for individual preferences and so thicker pieces can continue cooking in the pan while you're eating, and not be overdone. On the backstraps, cut the fat off the meat—this way you can remove sinew in between the layers and also slice the fat a little thinner than the meat, which makes it crisper and better flavored. Heat your pan—preferably a cast-iron pan—and then add oil. Don't fill the pan too full, especially if the meat has been previously frozen, because meat releases moisture, and too much water can cause it to cook "wet." Wet is when your steaks are bubbling in meat broth instead of sizzling in oil. Practice. Go ahead try it both ways; you'll notice the difference. If your pan goes "wet" halfway through, quickly heat a second pan and move some of the steaks to it for cooking.

Fry steaks the way you fried the liver and heart, but faster and for a shorter duration. Season, sear for a minute, turn to the second side, flip once more, and rush the frying pan to the table. It will keep cooking in the hot pan, so my advice

Cleaning and cutting a bull before I haul it home, with the Jade Mountains in the distance (Photo by Sarah Betcher)

is don't be polite. Eat fast! Keep your knife in your hand. Slice into each piece in the pan to have a look before you take it. How else will you know it's perfect? And don't forget to cook plenty of fat with your steaks. The fat tastes even better than the meat, and you're working hard, so it's good for you and will keep you warm in the cold.

Roasts should be cooked in a small amount of sizzling oil and meat juice, which forms in your Dutch oven or pot. Follow that one same rule: the meat needs to cook mostly in oil, with only a little water. This is simple to accomplish if you use a good pot, cut every roast so it has a bone, and season and sear it with a few chunks of fat in a small amount of cooking oil. Turn down the heat and let it cook slow. Each time you check your cooking, flip the roast over and peer in to see how much liquid has formed. If more than half an inch is in the bottom, tilt the lid for ten minutes or so to allow the extra water to evaporate. Practice. Make another roast tomorrow, and the next day. You don't want to ever go back to dry pasty caribou roasts.

Be aware that a high-meat and high-fat diet can cause you to lust for carbohydrates. Ice cream will be in your dreams. You'll prowl around the kitchen after dinner, searching for something sweet. Go ahead and put kindling in the wood oven, bring it up to temperature, make a quick batch of cookies or a cranberry pie. Cranberries help your stomach digest fat. Don't count calories. Those are numbers from a faraway place, and counting them will only make you hungrier. Out here, your body is tuned to burn fat and you are lean and strong. Tomorrow you'll head out again to gather more food, a small distant dot moving across the tundra under a huge sky.

FOLLOWING PAGES: *Lines of animals cross the sea ice as they migrate south.*

PART II

THE SECOND SEASON
OF THE YEAR

The morning is dark and cold. A huge frozen silence drapes the land, filling valleys and the tundra's undulations, and laying on soft pillows of snow under the skirts of spruce. The Darkness stretches in all directions, motionless and still, as far as the mountains and beyond. Overhead the stars glitter, twinkling as if they watched last night while the world, cloudless and unprotected, left the door open to outer space, and in the night that invisible spell poured down: *Cold!*

The cold of winter has returned. Summer buried the memory of winter until it was hard even to imagine, but now suddenly it is all so familiar—this cold that pinches like pliers, biting and merciless and dangerous. The air is harsh, brittle and dry, and searing, with most molecules of moisture wrung from it, condensed into frost. No insect exists in all the atmosphere of this world. The billion birds of summer have gone south. The few avian residents that remain have hidden in secret places, with feathers puffed against the long bitter night, surviving, or not.

Morning continues on, dark still, with the temperature continuing to drop. Finally, faintly the southern edge of the sky begins to brighten. The day is awakening, building a tiny fire beyond the horizon. Slowly the fire catches and throws a small orange-and-blue glow. Distant hills take shape, black against the dawn. The blue and orange begin to melt the blackness around the stars. Snowdrifts glow on the river ice, a suffocated blue. The land lightens, and in the north a purple curtain hangs above the horizon. The snow becomes a fairyland frosted in pale pink and purple and blue icing, and far to the south above the horizon where the sun is hidden, thin wispy clouds burn golden flames like tiny commuting dragons.

The first rays of the sun paint the mountaintops pink. The nearby tundra remains in shadow. Up on a ridge, a line of knobby tracks catches light, whiter than the surrounding blue gray, an old windblown wolf trail that wanders to the skyline and disappears over it. Windblown tussocks are a thousand dark heads where grass shows through the snow, with a thousand white drifts in the lee, and light reflecting off each scalloped surface. Far across the fling of tundra, stray scattered willows and clumps of alders dot the landscape, waiting in otherworldly silence, their branches brittle and bundled with frost. A faint puff of fog floats in the air, and one of the clumps moves. Dots take shape—a small group of caribou!

A few of the animals are standing, feeding. Nearby others are lying down, resting where they last fed. Their faces and flanks are frosty, their heads up and eyes watchful—watching for danger, and watching this winter day arrive—as if they too are breathing in the beauty of this morning.

A chip of sun is up now, beaming thin red rays that bathe the land in glowing light and shadows. The sun inches along behind the horizon but can climb no higher.

A lone caribou rises. More follow, unfolding their front legs stiffly. They stand scanning the tundra, alert and picking apart the landscape for possible predators. Finally, a bull begins to feed. Others follow, pawing at the snow with their front hooves and stepping forward to eye the exposed vegetation for the lichens they favor, quickly biting off the plants with the teeth of their bottom jaws pressing against their palate on top. The curved outer shells of their hooves have changed since fall and grown longer and harder and sharper to better break through hard drifts and layers of ice. Their eyes can detect a broader spectrum of light than other animals on the tundra, including ultraviolet light, which helps highlight the lichens against the white snow. The caribou swallow the food without chewing. Almost rhythmically a front hoof moves more snow aside. More vegetation is scanned, torn free, and swallowed.

Beside the feeding animals, others lie comfortable in the cold, ruminating, regurgitating and chewing the tough plants they swallowed earlier, grinding them into fine paste with their back teeth. Calves lie near their mothers with their forelegs folded under their small chests, patient, wide-eyed, and trusting, experiencing this, their first journey from birth in summer sun and now into the depth of winter Darkness.

Nearby the adult bulls are visibly larger than the cows, yearlings, and calves. The bulls are the least wary in the group. They are famished and feed steadily, greedily, and not always doing their share as sentries. The rut was hard on them. For the males, the month of October was one giant party. They ate little during that time, abandoned common sense and caution—posturing, fighting, chasing cows, and mating. Somehow most managed to escape predators and stay alive and keep up with the migration. The bulls have shrunken in the two months since autumn. Nearly all the wonderful fat they gathered over the summer is gone;

A red fox feasts on a caribou wounded and abandoned by hunters.

they are skinny now, with less meat on their bones, and less bone—their bones are growing thin with the limited diet and strain of winter—and those magnificent iconic antlers have fallen from the heads of the mature bulls, no longer needed and abruptly abandoned along the trail south in November. Without that weight and the accompanying hormones, their necks have shrunken, too. They've lost a third of their weight or more, and are vaguely goofy looking, with long heads on the ends of thinned necks. Even their attitude shows this loss. All their bravado of fall has faded. They eat quietly, deferential to the antlered cows. It's as if they wish they could explain: *Oh, if only you could have seen me in my prime!*

The cows have become smaller, too. Their faces are slender, their coats thicker— the hair on their back and sides has grown dark brown, tinged with white along the flanks, and their legs and hooves are fringed with silver. All the animals have thinner skin than in fall—weaker and harder to sew—and their hollow hair is thicker and in the process of growing more brittle. By late winter, it will break easily and shed. The animals no longer have the sleek look of autumn. Now their hairs point outward, puffed and fluffed to provide more insulation. Under the coarse hollow outer guard hairs, a second dense inner layer also traps air, providing additional insulation. The caribou have large hearts for their body size and make use of regional heterothermy to shunt cooled blood to their extremities, maintaining two internal temperatures: a high core body temperature and a much cooler temperature in their legs.

The pregnant cows still have their small hard brown antlers; these won't drop until calving time in early June. Non-maternal cows drop their antlers in early winter, and yearlings—both male and female, a year and a half old—have short antlers, roughly a foot or less in length, that they won't shed until March or April. At their throats, their manes hang shorter than the adults', and their faces appear longer than the fawns'.

Two- and three-year-old males, and even four-year-olds, haven't yet shed their small antlers and may keep them until April. From a distance the immature bulls are not easy to distinguish from the adult cows. Next year these young males will have slightly bigger antlers, with more curve—angled back and up and around—and will begin to take on the classic look of adult bull caribou. Now, in midwinter, these "teenage" bulls are the only caribou that seek regular interaction with each other, conducting mock battles, hooking each other with their small antlers. This causes the hair on their necks to appear pocked and gray from these play fights. This field marking on the males, and the black vulva patch on females—and possibly a calf trailing close at her flank—can differentiate a young bull from an adult female.

The calves have shed the velvet from their small antler spikes, and most have ceased nursing. They feed beside the adults, and roam farther from their mothers now that the fall migration is over. Here on the wintering grounds, small bands mix and mingle, somewhat sedentary, spreading out more than during the migration, with daily movement minimal unless weather brings a dump of snow, or danger appears. Predators—especially human hunters on snowmobiles—can cause the bands to flee in panic, covering miles in minutes.

This morning there is no rest for the cows and they must remain watchful, and again must eat for two—most are pregnant again and many also still have calves at their sides. And under the animals' skin, they all feed additional mouths—there a living terror has birthed and is growing larger, sucking protein from the caribou. Warble fly eggs, almost invisible when they were laid in summer in the hair of the animals' legs, have migrated up to the caribou's backs and burrowed in to feed directly on flesh. The larvae are already the size of maggots, and as many as a hundred or more hitchhike on each caribou. Their insect parents have chosen this warm and well-fed ride across the Arctic for these warbles, and for nearly a year they will eat and grow, eat and grow, until they become giant, writhing horrors burrowed under each caribou's hide. And in the nostrils of these caribou, the warbles' cousins—botfly larvae—are busy too, plugging the animals' sinuses with maddening, squirming, wriggling maggot colonies.

Caribou dig for lichens under the drifting waves of snow.

For now, in the stillness of midwinter, the animals' hooves squeak and crunch on the hard drifts as the caribou graze, steadily cratering down through the snow to feed. On the tundra the sun is gone, slipped below the horizon, and the day's sequence of oranges and blues and purples replays in reverse. To the south the fading orange glow stretches along the horizon, silhouetting stunted black spruce hunched like tiny figures hurrying home before night, and in the huge silence a lone raven passes overhead, black against the blue, the panting sound of its wings a sudden reminder of other lives inhabiting this frozen landscape. The raven rides in the sunlight above, and a bit of frost around its beak catches light, and glossy black feathers flash glints of orange. In a few moments the black dot recedes in the sky, passing from sight over a distant ridge.

THE COLD SPELL CONTINUES. EACH NIGHT GROWS LONGER, EACH DAY SHORTER. THE LAND holds its breath, waiting for what now is only a dream: light, sun, water, and warmth; birds singing; flowers, green leaves, and bugs buzzing—if those imaginary things still exist.

Finally, a night comes that is different: tonight the animals know something is coming. Foxes fling their raspy barks out over the moonlit snow. Lynx pad along the ridges, hunting, and occasionally screaming. Mink and weasels work their way across buried sloughs, ducking into the silent safety of caverns under roots and drifted cutbanks. Marten do the same in the spruce forests. Owls call from the shadows, and across the open tundra floats the howling of wolves.

A wolf pack roams the hills, moving down to the ice to follow a frozen river, until the scent of a young moose floats in the chilled air. Before morning the snow is trampled and the wolves have finished feasting. Miles to the west, a wolverine crossing steep ravines, searching for old bones and buried kills, catches the first hint of blood on the freshening breeze. The wolverine stops, scents the fresh moose kill, and lopes faster, hour after hour, tireless.

By morning, dawn has overslept and the moon and stars have vanished behind a gray blanket. A storm is moving in. The world is slow and sluggish to grow light, gray and dim, and the day barely begins before it is getting dark again. Wind gusts in the willows. Grains of snow slide along the drifts. The first snowflakes slant down.

On the tundra, the band of caribou plods east, facing into the frigid wind, trusting the moving air to warn them of danger. In the dimness their eyes make use of ultraviolet light to spot the glow of predator urine in the snow and any pale furry wolf-shaped forms lurking nearby. Undeterred by the wind, the caribou paw at the drifts for forage, their dark shapes disappearing and reappearing in the storm. Snow coats their faces. In addition to conserving heat with their twin layers of fur and dual thermostats, the animals make use of their long noses and that extra surface area of their sinuses to preheat the frigid air and extract moisture. They conserve additional water by urinating less and recycling urea within their bodies. Some of the animals stand with their butts to the wind, appearing to be oblivious to the searing cold, while others face into it as the drifting grains strip the tundra down to a crust of ice. Dwarf birch twigs protrude from the snow, waving in the gusts and sending their last seeds sifting away on the wind. Small spruce trees lean forward in the gale like forlorn children all carrying the same sadness.

Slowly the caribou move toward a line of trees. Individual animals stop to paw through the layers of ice for lichens. Snow fills their craters. Finally, they lie down to rest and chew, and sleep, seemingly unfazed by the howling storm, the bitter cold, and the snow that drifts into crescents around their buried bodies.

FOLLOWING PAGES: *As the snow grows deeper, travel on foot in areas protected from wind requires snowshoes. Sliding grains of ice cut away the snow, exposing vegetation on tundra, windblown ridges, and mountains. The wind forms sastrugi—drifts so solid that animals hardly leave a track.*

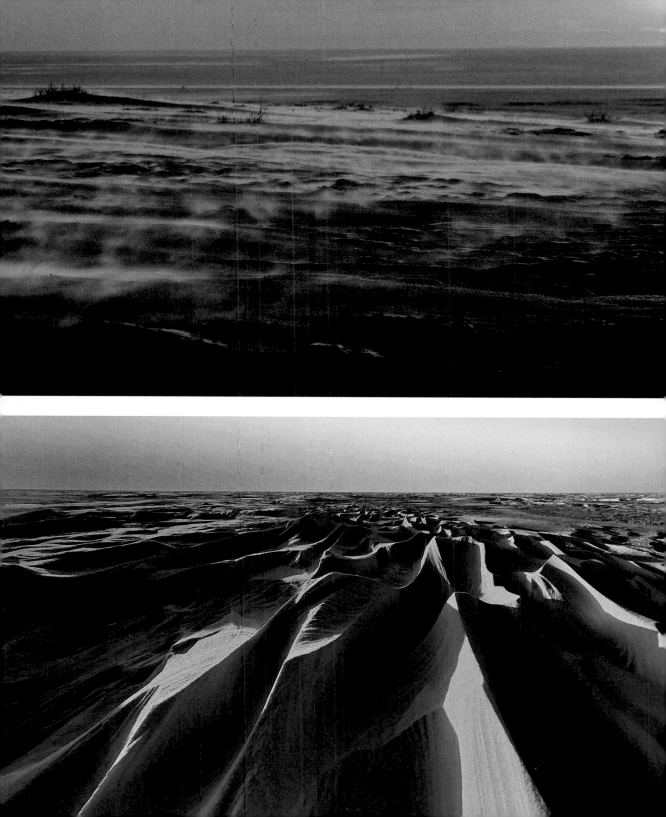

CHAPTER 7

CARIBOU COMPANIONS

On the river ice my dad stops walking. He's wearing baggy drab green wind pants and a blood-and-pitch-stained parka with a wolverine ruff. Slung on his back is his .270 rifle. Across his shoulder he carries a double-bit ax. My brother and I stop a few yards behind. We're six and seven years old, and we know well not to crowd together on new ice.

The river has finally frozen enough to walk on, and it fills our chests with wonder and exhilaration to be able to walk on top of water again—after the summer of bogs and bugs, and green leaves growing everywhere collecting sunlight. Now we're moving back into winter, and we are winter people. We've all been shuffling our mukluks, to keep from falling on the black glare ice slippery with fresh white powder on top. Behind us three trails show where we left the grassy bank in front of our sod igloo and started upriver. We're out scouting for firewood. And meat. Our family always needs both.

We boys stay silent, glancing at the side of my dad's face, to see if he's thinking the ice might be unsafe, or why he stopped so suddenly. We follow his gaze. He's not looking at the ice. He's listening, and watching a raven circling in the sky.

An ermine peers out from among artemisia stalks.

Two caribou suddenly appear from out of the willows. Caribou have been migrating through for two months, but we freeze and go hushed, watching the animals. They have their winter coats now, with longer, thicker hair and pretty white markings, and their hair is fluffed up in the cold. Their hooves clatter across the frozen gravel above the mouth of Amaktok Creek. The two animals leap over heaped ice pans along the shore and trot out onto the river ice.

"Cow and calf," Howie whispers.

On the ice, we all simultaneously bend our knees, gradually folding down to the snow so as not to stand out so bold and black on the glaring white. "We're not trying to get them, right?" Kole asks.

"No. I think—"

Out of the trail in the willows where the caribou appeared comes a wolf. It bounds across the gravel and sprints onto the ice. The wolf is black and looks large on the flat white expanse. On the slippery snow-covered ice, the cow caribou falls. Suddenly her calf falls, too. The wolf races toward them.

"Are you going to try to shoot the wolf?" I whisper. I'm hopeful that he'll say yes, but Howie shakes his head. "Just watch."

The wolf sprawls on his side. The caribou scramble to their feet. They run straight out across the river, racing for the safety of the far shore.

The cow falls again. The calf splays out beside her. The wolf is close behind. His front legs slip sideways. His throat hits the ice. The cow gets up, hobbling. Her front leg flaps, the bone terribly shattered. The calf stops, and then runs down the river ice, coming toward us. Suddenly the wolf senses danger. It turns to stone, its ears up and the pointed black silhouette staring at us. The wolf swivels and flees toward shore, racing up the bank, vanishing into the willows.

My dad sighs. "Agh. We interrupted here. Wolf had that one. He won't be back now." He pulls his rifle off his back, loads the bolt, and drops the running calf. The boom of the gunshot echoes back from the timber along the far bank. Kneeling on the snow, he swivels, his knee sweeping snow aside, exposing black ice. His next shot finishes off the wounded cow.

I scoop up the warm brass cartridges and shake them dry. Without thinking, I hand one to Kole, sharing as we always do. Both of us are watchful and waiting.

As the soil under the tundra thaws, white spruce are growing rapidly, which can affect drifting patterns and the winter snowpack.

Quickly we dig under our snowpants to our inside pockets for our sharp little knives and follow Howie, already moving toward the larger caribou. When we get to the animal, we roll the carcass on its back. Kole and I hold the front legs, balancing it while we examine the shattered leg and protruding bone. Howie makes cuts—first a few quick slashes to cut away the mammary area and a little skin around it, exposing the taut sinewy abdomen. He flings the flap away on the ice for foxes and ravens. Drips of leaking milk mix with a little blood and hairs. The calf had been nursing. For just that moment, feet widespread, our dad pauses. Gripping his knife, he scans the riverbanks, upriver and down, swiveling his gaze along the mile east and mile west of big white ice. He inhales, and bends down again to his work.

My brother and I pull off our mittens and lean in with our knives. Inside the skin, the animal feels hot and good on our fingers. Already Howie has the brisket bare and is making a small slit in the abdomen. With two fingers, he pushes the warm slippery stomach in, enough to insert the tip of his knife without cutting the guts. He runs the cuts up to the brisket and in the other direction to the pelvis. The big gray-green stomach swells out. Howie's knife plunges into the joints between the forward ribs and brisket, and from inside the chest cavity the lungs make a loud whoosh. He cuts each joint and swings the brisket open. His knife flicks, gesturing for us to continue skinning. "Open the neck." He straightens up and scans the ice again for any movement of animals.

Kole and I make cuts from under the skin—cutting from the outside would release handfuls of hair—and Kole slices along the throat to the jaw. Howie severs the windpipe and throat behind the tongue, working back toward where it enters the chest, lifting and pulling, until the lungs and heart are out. He tears the diaphragm free and gently rolls the bulging stomach to the left side. A few more cuts at tissue and the whole pile slides onto the snow.

Ravens pass overhead, cawing their approval. They travel on upriver, not yet coming close. Kole and I cut along the insides of the jaw and remove the tongue. Afterward we go to the rear and clean out the pelvis. Howie flips the carcass over to drain. He steps to the gut pile. "Boys? You want the bible?"

"Yes!" we shout.

He rolls the stomach over and cuts free a large kidney-shaped section. Boiled, this third stomach of the caribou, the omasum or "bible," is a local delicacy. He slits it open to reveal the many layers, all packed with dark gray-green stomach contents. He scrapes it clean with snow, and then scrubs his knife and fingers with more snow. For now we are done.

Quickly we cross the ice to the calf and do the same. Howie leaves the second bible because it's not fat, and again we scrub our bare bloody hands and knives with snow. Our fingers turn red and rubbery, cold and wet. We shake off the pink slush, wipe them clean on the calf's soft fur, and slide our blades back into our sheaths. It's all taken just a half hour, this unplanned event in our day. Howie loads the V-shaped briskets like trays for us to carry the tongues and kidneys and livers. He gathers up the hearts, *itchaurat*, and bible.

Once more we eye the sky and the ice for predators, for food, and for companions, and then turn and hurry home for a sled.

WHEN WE RETURN, WE LEAVE OUR WOODEN SLED AT THE FIRST KILL SITE AND WALK FARTHER up the ice toward Amaktok to examine the intersecting footprints of the caribou and wolf. My dad has taught us to enjoy reading the stories left by animal tracks and to harvest the companionship they provide. He's an expert in the language in the snow, and we watch his eyes scan back and forth. Living alone on the land, out so far from people, we are extra curious about our neighbors here, the animals. In the loose powder, the tracks are indistinct, just streaks where the black ice has been exposed, with larger areas of pushed-aside snow marking where each animal slipped or fell. Finally, we find a few drops of blood where the cow's leg snapped. We backtrack to the shore. Howie points out perfect caribou hoof marks in overflow slush. He finds wolf paw prints, fat pads pressed into the snow, and points out where a toe is missing on the wolf's front foot. We search for more tracks to double-check; sure enough, only three toes show on the right front foot. We speculate and decide the wolf likely had been caught in a steel-jaw trap.

We start back down the river. A string of fox tracks loops in and out of a thicket of willows. Clumps of grass protrude from the snow, tawny in the low light of the October afternoon. Tiny dimples show where voles and shrews have traversed the drifts, and round tunnel entrances mark where they ventured out from the safety of trails under the snow.

We spot no tracks of any other animals—no sign of people, no human sounds in the air. The black dots of ravens cross the sky, eyeing our fresh kills, steadily circling closer. For a moment, Howie and Kole and I stop, and listen. In the silence from underfoot comes a sudden eerie boom, the river ice settling. The huge reverberation is scary, thrilling—like standing on a sleeping giant with an upset stomach—and a galactic sound streaks through the ice, echoing to the far shore. We smile in awe, still listening to the land around us, huge and silent. We breathe again, and angle back to the sled.

WHEN WE GET BACK TO THE DOG-FOOD CACHE WITH THE TWO CARIBOU, MY MOM WALKS down to the shore with her *ulu*. The four of us gather around the larger animal. Ravens circle the gut piles upriver, and more ravens call encouragements from the tops of tall spruce. We cut the neck off the cow caribou. The leggings are nondescript brown—not pretty skins for mukluks—and one leg is torn by the shattered bone, so Howie tells us to go ahead and cut the legs off at the ankle and wrist joints. The cow's hooves are hard and long, already in the process of growing bigger and sharper for the long cold season of cratering down through snow for food. Kole and I cut the back legs, leaving the Achilles tendons intact for handles to carry the meat by. We are careful to find the third, lowest joint; locally, not doing so is a sign that you don't know what you are doing as a hunter.

When we are done, Kole and I slide on the slippery ice while we await more instructions. My parents talk it over; they decide to leave the cow whole with the skin left on. The meat won't freeze right away and will have a little time to age. And during the winter, the skin will protect the meat from drying out in the wind and will insulate it from seasonal temperature fluctuations. Our freezer is the outside air and a caribou's hide the best freezer wrap. Winters are a deep freeze and the meat will keep till spring, or until we eat it.

The calf has beautiful leggings with silver fringe, as calves often do. We admire the hair and know what to do. Skinning leggings in the evening was one of our first chores as small boys, one that earned my brother and me a quarter for each legging. Now we slice up the front of the front legs, the back of the back legs, and carefully down between the knuckles and around the black hooves, keeping all the skin and fur and even bits of black hoof past the last wiry white hairs. The entire legging is needed when matching sets and cutting skins to a pattern—this my parents have been taught by Iñupiaq seamstresses—and must be done properly.

Howie skins the calf rapidly and cleanly, doing most of the skinning with his fists so the hide will come away perfectly. When he's done, Kole and I grab the furs and run fifty feet out on the ice—far enough offshore so overflow water won't freeze

High noon in winter shines through frost formed in the night. A red squirrel gathers alder seeds in the cold.

the skins in—and help my mom sweep fresh snow aside. Mama stretches the warm wet calf hide and leggings and sticks them down flat on the cold ice. Tomorrow we'll peel up the flattened hide and leggings and hang them to dry behind the igloo in the north wind. Cold, dry wind makes skins cure white and strong—best for tanning.

The calf is too young and small to save sinew off the back for sewing thread, and we quickly quarter the animal and cut the ribs and backbones into sections. The meat is not fat but will be tender and good to add to fatter meat for soup. The only fat is a thin white rim around the top of the heart and a small amount on the brisket. My mom points this out. If this were a big bull—hunted earlier in the fall before the rut—the lower leg bones would be kept for marrow and soup, and she'd be saving *itchaurat* and kidney fat and what we call poop-chute fat from the lower intestines, all kept separate to be rendered on low heat and poured into jars and used later to make *akutuq*, and the harder back fat saved for pemmican.

We sled the cow up the trail to store with a dozen other frozen bulls on a flat log cache built seven feet off the ground to keep the meat safe from wolverine and foxes. The posts have fuel-can metal nailed around them to keep shrews and voles from climbing up to feast on and tunnel into the caribou, eventually leaving only a hollow shell with a turd-dotted skeleton inside.

There are fewer caribou heaped on our cache than just five years previous. Now we have a stack of frozen whitefish, sheefish, and trout, too, all signs of many changes in our lives. In the past few years my parents have acquired fishnets and a gas-powered chainsaw, and my dad gave away most of his dogs and bought a Canadian snow-traveler, a strange red metal creature with a one-cylinder engine, two tracks in the front, and wooden skis in the back under a bench seat. It no longer takes my family two days to get to the village, and along the trail we no longer carry caribou for dog food. After Howie's favorite lead dog and some of his other dogs died, he decided to buy the snow-traveler. I don't know all the details and contributing factors, but apparently relying on getting enough caribou each fall to feed our family and his dogs had been hard on my dad, and midwinter dog team travel with infants, and later toddlers, had been challenging for both my parents.

In just a few short years, my dad has gone from having to kill nearly a hundred caribou each fall to instead getting only a couple in August, eight or ten in late

September, and one or two during breakup. Caribou still make up a majority of our diet, but the constant stress of keeping a dog-food pile through the winter is gone—for now at least. Now he often travels on foot, or on snowshoes or skis, to check his traps and to hunt. When it's time to haul loads of firewood and our snow-traveler won't start, we all gather around, silent, somber, and supportive, the same way we do when his reloading press has jammed. We don't think deeply about how the widespread embracing of machines into a culture of hunter-gatherers might change the course of our future. We just worry that our red machine will be broken and our house will grow cold.

The work is done, and we scrub our fingers and knives with snow a last time and go inside for lunch. Our sod house now has a door made of spruce boards and caribou hides. Howie buried the old tunnel a few years ago, and shoveled aside dirt of the hill to add onto our home. Inside at the stove, Mama fries the liver and *itchaurat* and slices from the top of the cow's heart, the part with fat on it. She puts the brisket and tongues on to simmer for dinner. Kole and I put the kidneys and bible in a small pot to simmer, and then we bring the cow head in to skin on the table. When we're done, Kole and I saw open the skull. Mama gives us a bowl to put the brains in to fry for breakfast. We beg her to cook the head, one of our favorite meals, and she agrees—when the roaster pan is empty, after the brisket is gone. Today we still haven't scouted for firewood. We didn't make it far up the river ice. It doesn't matter. Each morning brings a new sky, and each sky determines what we do with the day. We have a big woodpile, and hauling logs home can wait for another tomorrow.

ONE AFTERNOON A FEW DAYS AFTER GETTING THE TWO CARIBOU, MY DAD RETURNS HOME on snowshoes to tell us of spotting the black wolf on the tundra. He says he thinks it's a resident—as opposed to the traveling wolves and wolverine that pass through trailing the caribou migration. Again, he didn't try to shoot the wolf. I don't ask how the rest of my family feels, but I'm disappointed. I wish he'd have shot the wolf—to skin it and have the hide be ours; to count it as something gathered, gotten, owned, and to later possibly trade or sell or sew with. A wolf pelt hanging in

our house feels valuable, a sign of ability, strength, and pride. Disappeared on the tundra, it feels like nothing, an opportunity gone like smoke in the wind. We're a family of hunters. The Iñupiat Eskimo people we know and interact with are hunters and gatherers, their culture focused on the land and animals. Shooting a wolf is what people do, and not shooting one is odd. Our dad is an odd man, though. He doesn't ever brag—especially about killing animals. He knows well the harsh conditions through which every creature here struggles to survive, to find food and somehow raise young. And he is awed each day by the Arctic and quietly grateful to live in the wild, and part of his eccentricity is the way he remains reverent of the company of the nonhuman hunters we share this land with.

All this is in our small igloo, unspoken. All this is confusing to my young and limited comprehension. I only know that I want to be a hunter, and hunters don't let wolves go. Illogically, I also want to be Eskimo, which feels like the only way to claim to be part of an elusive community, authentic, normal, and connected to a long, local history, with hunting at its core. In this way, in these young years, I'm simpler than the rest of my family. The careers of the Outside lands don't enter my dreams. Hunting is the clearest trail I see and the one I will follow as far as I'm able to.

Mama doesn't say how she feels about Howie letting the wolf go. Kole doesn't say either, and we kids don't see the animal again. The snow is getting deeper, and we boys are clumsy on our small snowshoes; we can't go as fast or as far as our dad. The caribou migration is nearly over. Only a few stragglers are left now and soon the last fading forms will disappear down the drifted trails, moving south toward their wintering grounds. The land feels lonesome and empty after their immense passing. They will be back in the spring, of course, at breakup, that most exciting time of year. But spring is way off on the far side of the longest coldest season. Ahead lies a land that will be reshaped again and again by the storms of winter, until it is a completely changed world that we walk upon, with this season buried and left behind in a distance made of cold days and Darkness.

A wolf searches the sea ice for food. A lynx's pads compressed the snow, which then hardened, leaving inverted tracks when the wind blew away the softer, looser snow.

CHAPTER 8

THE NORTH SLOPE

My parents were raised in Ohio with cars and roads, restaurants and theaters, churches and bars—that stuff people sometimes refer to as the "real world." They didn't know each other there; cities are large, and they lived in different ones and were raised under different religions, too. They came separately to Alaska for college, Howie in 1953 and Erna in 1960. In that dimly detailed past, apparently Howie first traveled to the Arctic to work as a caribou biologist's assistant. Later, when my parents came here to live at Paungaqtaugruk, they brought dog harnesses and nails, a keg of gunpowder and other supplies, and something else—something that got stirred like salt into the soup of my igloo-boy upbringing. That thing they carried was science.

When Howie talked of the caribou research work he had done, ironically he most often mentioned how fast he could run, chasing down newborn calves, for science. That description, and the thought of my dad with an actual career, seemed incongruous to me as a kid, as did the reason for the research: Project Chariot, a scheme by the United States government to use atomic bombs to blast a harbor to the west of here, along the Chukchi Sea coast.

My parents had lived the Chariot years—a politically tumultuous time in Alaska—in and around Fairbanks and had been friends and

On the Utukok Uplands, a cow leads her calf away from where it was born a few hours earlier.

coworkers with many of the scientists involved in the pre-blast research. The tales they told us of that shadowy project as we sat around our family's kerosene lamp were as warped and weird as the chapters Mama read aloud from *The Thousand and One Nights*—except my parents insisted these stories were true.

Like the accounts we heard from Eskimo hunters of no caribou here along the Kobuk, the Chariot details weren't chronological and seemed nonsensical, almost like fables, and once more I listened and stashed that fantastical information somewhere between curiosity and disbelief.

AFTER WORLD WAR II THE UNITED STATES GOVERNMENT WAS BUSY BUILDING AND TESTING nuclear weapons, but with one major component missing—an enemy large and wicked enough to drop those bombs on. The generals, politicians, and scientists involved needed to justify their new creations, and, like most boys, they loved to blow shit up. For these reasons the US Atomic Energy Commission (AEC) invented the Plowshare Program, a progressive and pleasant-sounding government program to, well, blow shit up.

In those years the northern part of the Alaska Territory was considered wasteland, a vast barren desert of tundra, ice, and snow. To the men in charge of Plowshare, it was perfect. What could be more brilliant than nuking the Arctic in the name of progress? Before long a dot on their world map became Chariot, a site termed Ground Zero, where the initial plan was to vaporize a coastal area with a string of atomic bombs and a larger hydrogen bomb or two to create an instant deep-water port.

On the actual earth, that pinprick was the mouth of Ogotoruk Creek on the Chukchi Sea coast, a hundred miles above the Arctic Circle, between the Iñupiaq villages of Point Hope and Kivalina. There, a small creek flowed into the sea, and waves lapped against a pebble beach. Directly to the north rose the cliffs of Cape Thompson, vertical rock faces plugged with the nests of tens of thousands of seabirds, the air sharp with the smell of guano and ringing with the shrill cries of murres, kittiwakes, puffins, gulls, and other birds. Offshore, in the shallow ocean between Alaska and Siberia, the Chukchi teemed with fish, whales, walrus, seals, and other sea mammals, harvested seasonally by subsistence hunters from nearby villages. To the east, the tundra, hills, and mountains stretched for hundreds of miles toward the distant border with Canada—land now included in the National Petroleum Reserve in Alaska (NPR-A) and Arctic National Wildlife Refuge (ANWR)—and known even then by private industry and government agencies to be underlain with vast quantities of coal, oil, and minerals.

Proponents of the project claimed Iñupiat Eskimos would benefit from the "development" this new port would bring. Alaska business leaders, church groups, and politicians—including the new state legislature in 1959—joined a growing chorus clamoring for this fantastic and futuristic project. Scientists and university professors and others who warned against the project were ostracized, and in some cases fired and blackballed from future employment in the United States. Already Project Chariot was a pipeline of politics and money flowing north to the young state of Alaska.

Not far from the Chariot site, across nearby low rolling hills and valleys, were the calving grounds of the Western Arctic caribou herd, the largest remaining herd of migratory land animals in North America (not to be confused with the well-known Porcupine herd that migrates to the coastal plain in ANWR). The population was in recovery, rebounding after the better part of a century of decline (a decline arguably caused by the introduction of a previous weapon of war: rifled guns).

In large and small groups, this immense herd wandered the windswept land, migrating with the seasons, traveling vast distances—individual animals traversing as much as three thousand miles annually between breeding, calving, and wintering grounds—surviving countless river crossings, ice breakups, blizzards, threats of starvation, wolves, winter-dark days, horrific parasites, grizzly bears, human hunters, and hordes of insects. Thousands of wild caribou living daily lives daunting to the imagination, while far to the south, the enormous wagon of human history was again veering on an uncanny path straight toward their homeland.

AT THAT TIME, IN THE LATE 1950S, A YOUNG STUDENT NAMED PETER LENT HAD JUST GRADUATED from the University of Alaska with an undergraduate degree in biology. He lived

in a small cabin in Fairbanks and was interested in small mammals—ground squirrels, lemmings, and voles—but also was toying with the idea of continuing on to grad school to study whales. He—presciently—believed whale research would become a big deal someday. He'd never seen a caribou. Fate, and Project Chariot, were about to change that.

Peter Lent was not the only scientist to be suddenly swept up in the huge and rushed research effort demanded by the AEC. Scores of anthropologists, biologists, and other scientists were hired with AEC funds diverted to the University of Alaska–Fairbanks and other institutions to conduct extensive studies of the flora and fauna and the indigenous people living near Ground Zero. A Scandinavian reindeer specialist was offered the job of surveying the Western Arctic caribou herd. When he declined, Peter got the job. "I was twenty-three, just out of college, and running with very much freedom a project with a budget of $100,000! And a stupendous salary of $6,500 per annum."

It was a golden opportunity to study a large and interesting animal in a far-off and expensive-to-access area of northern Alaska. Peter soon found himself spending weeks, and then months, putting in hundreds of hours flying in small aircraft to try to keep track of the movements and distribution of caribou on the Arctic landscape. He needed assistance on the ground, too, for the collection of caribou samples for radioisotope studies.

For this field research, his assistant would need to collect five caribou a month for samples, and also to accompany him to the calving grounds of this legendary Arctic herd. Peter's plan was to follow the herd on their annual cycle, even to be dropped off by bush plane prior to spring calving in a remote area north of the Brooks Range. For this he needed a companion he had the utmost trust in, one who would remain calm and capable—even cheerful and humorous—while completely isolated and facing constant hardships, including months of traveling by small dog team on snow and by foot after the snow melted, crossing miles of soggy tundra in the more or less constant company of grizzly bears, swarms of billions of biting mosquitoes, wind,

Howard Kantner with his lead dog Chitina. Howie weighs a newborn calf, June 1961. (Howard and Erna Kantner Collection)

rain, summer snowstorms, and those science-fiction-like nights under an unrelenting sun. Peter chose a close friend—a graduate in zoology from the University of Alaska—named Howard Kantner, known to his friends and later to his sons as Howie.

The two friends had winter camped and beaver trapped, had built small cabins, and separately and together had conducted biological research in the Aleutians and Southwest Alaska as well the Southeast and Interior regions of the Alaska Territory. They had hunted moose and other animals, and Howie had camped and hunted caribou around Fairbanks and McKinley Park from the Fortymile, Delta, and Nelchina herds. Resident caribou in such areas, though, were a somewhat different beast, boreal forest and mountain caribou, much more sedentary than Arctic caribou, larger—the bulls weighing four hundred to six hundred pounds—living in much smaller groups, and seasonally traveling much shorter distances. The near-mythic herds of the Arctic were colossal in comparison, far to the north, and forever roaming a landscape almost too large to comprehend, with few human inhabitants, and countless unknowns.

One theory among biologists at the time was that Alaska probably held six major caribou herds, scattered from the Arctic to the Alaska Peninsula, and five minor herds—most of those in the Interior. The true number of herds in the Arctic—a herd being defined as caribou using the same range—hadn't yet been established. Estimates listed the total caribou population in Alaska at 1.3 million animals, an educated guess, further compromised by the fact that the Porcupine herd roams back and forth across the border between Alaska and Canada.

Worldwide, caribou populations were rebounding and were believed to be around 5 million animals, with herds more or less everywhere in circumpolar lands, and stretching across North America—mostly in Canada—and even occasionally down into the United States as far south as Idaho, Washington, Vermont, and other states. Biologists believed that many different species of caribou existed—barren land, boreal, mountain, Peary, reindeer—when in fact they are all the same species and can interbreed, given the opportunity. These caribou instead make up seven subspecies, bearing the signs—especially in the case of the diminutive Peary caribou—of extreme evolutionary adaptation over eons in the harsh climate of the Arctic.

IN MARCH 1961, WITH FIVE DOGS, A SMALL SLED, AND A HEAP OF GEAR, PETER AND HOWIE flew to Kotzebue, a small Iñupiaq community above the Arctic Circle on the Chukchi Sea. They stepped off the plane onto the packed snow of an airstrip behind a hospital, two churches, and low school buildings. Between the structures, the white sea ice was visible, stretching out to the flat line of the horizon. Small cabins and tarpaper shacks along the shore were buried by wind-sculpted snowdrifts, the only demarcation between land and ice.

Peter knew his way around the town. He'd spent the better part of the previous year there flying surveys out of Kotzebue, with plenty of time stuck on the ground waiting out bad weather. He led Howie over the drifts to a cabin hidden in the shadow of the two-story Drift Inn at the far end of Front Street. The men slid down a snowdrift and ducked through the outer door of the cabin. Inside, the entryway was crowded with leaning shovels, axes, wooden kegs, buckets, rolls of sealskin, strings of hanging dried whitefish, and other assorted possessions tucked out of the weather. The pervasive smell of rancid seal oil filled the *qanisaq*. Peter knocked and opened the inner door.

A blast of warm stuffy air enveloped the men. Inside, the single room was sweltering. Along two of the walls were two beds. Behind the door was a brown drip-oil stove, and beside it under a square of cardboard was a chamber pot. Under the window on the south wall, a small kitchen counter was heaped with dishes and pots, jars of dark liquids, a spattered sourdough container, a box of Lipton tea bags, and cloth bags of sugar, flour, and noodles. Sitting on a bed was an elderly Iñupiaq woman, tiny and thin and wrinkled.

She shouted and leaped to her feet when she saw Peter. "Ah! You come!"

Peter introduced Howie to Mamie Beaver. She stood smiling up at him, and turned to stare fondly at her friend Peter. She hurried to a Coleman stove and pumped rapidly, heating water for tea. She glanced over her shoulder. "You hungry? You boys eat?" She rushed around putting out crackers, tea, and *niqipiaq* (Eskimo food).

Howie liked Mamie immediately. She became a close friend of his, and two years later my parents were married there on the shore in front of her one-room house, and she remained a friend to our family until her passing in 1987.

Peter Lent in the field in the Interior of Alaska (Howard and Erna Kantner Collection)

After lunch Peter walked down the drifted shoreline to arrange for transport farther north with his friend Nelson Walker, a local pilot. Nelson was a young adventurer from Missouri who had built a rough and rewarding life in Kotzebue as a bush pilot, a hunting guide for polar bear hunters, and owner of a liquor store. Nelson had "married in," as folks termed it, to a local Iñupiaq woman, and together they raised a family in Kotzebue. Nelson had a sharp wit, a sense of irony, too; he named two of his sons after his most lucrative merchandise: Johnny and Hiram Walker, and also came up with the lasting nickname "Caribou Pete," as Peter Lent was later known in Barrow, Kotzebue, and villages across the north.

That night the men slept side by side on the few square feet available on Mamie's floor. Along the wall her young son, Eddie, slept on the second bunk. The following day was cold and clear. Nelson flew Peter one hundred miles farther north to Cape Thompson, where the AEC had built an airstrip. Howie arrived the following day with more gear. The sled and the dogs in crates were transported north by Wien Air Alaska with a larger aircraft.

When the drone of the last plane faded to the south, Peter and Howie stood surveying an open landscape that stretched to a horizon of sea ice to the west and up the cliffs of Cape Thompson to the north. To the east Ogotoruk Creek forked into low rolling tundra hills. The men chained the dogs to willows, and began sorting their gear.

IN THE FOLLOWING WEEKS, PETER SPENT MOST OF HIS TIME FLYING SURVEYS, WHILE HOWIE struggled to follow small groups of caribou over the hard drifts and wind-scoured frozen ground—bare ground in the Ogotoruk area being the reason caribou often overwintered there. Howie was impressed with the hard snow and good walking, after the soft, deep powder common around Fairbanks where he and Peter had spent much of their time. He marveled at how snowshoes were unnecessary, and used his small team to make forays up Ogotoruk Valley to make further observations. The terrain was new and awe-inspiring after the confining overgrown country of Interior Alaska, and each morning Howie woke up excited to set out across the tundra, reinvigorated even though the work was hard and tedious, and it was virtually impossible for him to keep up with animals so adapted to the landscape, so fleet-footed, so relentlessly on the move.

The long days of March grew into the longer days of April, and the fierce sun glared off the snow-covered land. Nelson Walker returned to fly Howie east to join Peter closer to the calving grounds. By May there was no more darkness, not even twilight. Peter and Howie were deeply tanned in the faces and hands; they were lean and young, tough and tireless, and could run for long stretches on the wind-hardened snow. Staying abreast with the moving caribou groups was short-lived at best, though, with no possibility of keeping track of individual animals. No radio or satellite collars existed, and the men could do little more than count animals, record the makeup of mixed-sex groups, and speculate at where they had traveled from and were vanishing to.

In their journals Peter and Howie marveled at how effortlessly the caribou crossed the uneven terrain, and how quickly they disappeared into the landscape. The men saw almost no sign of wolves, most of them having been killed by aerial bounty hunters—often making use of unlimited aviation fuel left behind in drums by oil exploration teams. Nowadays it is known that caribou can achieve speeds of nearly fifty miles per hour and swim six miles per hour—outpacing both grizzly bears and wolves on land and in the water—but there on the tundra in the spring of 1961, the young scientists paid dearly in hours and days and calories for tidbits of slowly gathered knowledge.

Every five days or so, they shot a caribou, weighed and skinned it, and conducted a careful necropsy: noting the condition and general health of the animal, including visible fat placement, and if females were carrying a fetus. They cracked a femur to check oil content in the marrow and quantified any parasites, including warbles and botfly larvae; tapeworm cysts in the meat and liver; brucellosis in the joints and synovial fluid; and echinococcal cysts in the lungs. Samples were placed in tiny glass vials. The vials were stuffed under their parkas to keep them from freezing and later shipped to Fairbanks, where Peter would study them in the coming winter months. Other samples were sent directly to the Lower 48 for radioisotope analyses, and samples of caribou milk were carried by Nelson to Kotzebue where his son Hiram was in charge of keeping the milk refrigerated until Peter could ship it on.

Years later—even now, sixty years gone by—I occasionally find a glass vial in the leaves under my dad's log cache. The floor of the old gray cache is made of rough-hewn spruce poles, and inside are Howie's old wooden junk boxes, full of random treasures—bent spikes, chainsaw files, scraps of smoke-tanned moose hide, logging wedges, greasy used sparkplugs, and soft cloth sacks of empty .270 brass. Occasionally small items fall through the cracks in the boxes and drop between the spruce poles to the ground. Those old spare vials are smoky with dust and age, and the miniature steel caps are rusted, squeaking when I unscrew them.

WHEN THEY WERE DONE WITH EACH NECROPSY, HOWIE SLEDDED THE CARIBOU BACK TO camp for meat to eat and to feed to their dogs. Ravens circled overhead, eyeing the humans, waiting for gut piles they knew would be left behind. Nearby, grizzly bears and other predators were sampling the herd, albeit for different reasons,

In June, cotton grass flowers look like snow on the tundra.

and the men took notes and parasites from their kills too, and from various feces left behind.

At that time, radioactive contaminants from the testing of atomic bombs were suspended in the atmosphere, circling the globe. Air currents and testing of nuclear bombs in Siberia were providing the Arctic with more than its share of "fallout," that fashionable new word coined to describe particulates from atomic detonations. Peter was especially interested in the tundra lichen commonly known as reindeer moss, or *Cladonia rangiferina*. The plant has an astounding ability to absorb radioactive isotopes such as Cesium-137 from thin air. The lichen also happens to be the caribou's favorite food and prime source of easily digestible carbohydrates.

Caribou are ruminants, with four compartmentalized stomachs and rumen muscles to send their cud—the vegetation they swallow without chewing—back up to be chewed at a later point. This intense chewing done by a caribou's back teeth grinds the cell walls of the plants and allows their final digestion to wring out the maximum amount of nutrients. In the case of caribou and *Cladonia rangiferina,* that digestion process was transferring high quantities of Cesium-137 into these wild Arctic animals. The radioactive isotope was then stored in the caribou's fat and further transferred to the rich milk that the cows produced for their young, passing the isotope directly to their newborn calves.

Research across Greenland, Canada, Alaska, Siberia, and Scandinavia in the 1950s and early 1960s showed something else that caused biologists even more concern: predators such as bears, wolves, wolverine, and other animals—including foxes, eagles, and other birds and creatures that fed on caribou carrion—were not only absorbing the radioactive isotopes like the caribou but their bodies were bioconcentrating it to exponentially higher levels. This was happening in the oceans too, where small fish ate plankton and other water-filtering organisms, and bigger fish including salmon and Dolly Varden trout ate those fish, and along the coastlines seals ate the bigger fish. Grizzly bears ate caribou and fish *and* seals. And the ultimate predator, humans, ate the caribou, fish, seals, and bears.

IN LATE MAY OF 1961, NELSON WALKER LANDED IN WHAT HE CALLED HIS FAMILY CRUISER, a four-seater ski-plane, and took the huskies south for the summer. He returned with his smaller two-seater Super Cub on oversized tundra tires to ferry the scientists' camp farther north. From there Peter and Howie continued following the caribou on foot. In the first days of June, they were able to observe the colossal calving event when tens of thousands of female caribou of the Western Arctic herd give birth more or less simultaneously on the Utukok Uplands. It was Howie's job to catch newborn calves, a few hours after birth, before they could run too fast. He quickly weighed each, noted the sex, and placed a tag in the calf's ear before letting it run back to its mother. To this day, Howie shakes his head when talking about the job—not judging, not bragging, definitely not impressed—only observing the irony and dichotomy between how much he loved being out there and how little their efforts likely accomplished. "I think only one tag was ever returned," he said, "from a hunter in Point Lay. Or maybe it was none."

In early June the sun shone night and day. The tundra was mottled brown and white with melting snowdrifts and protruding tussocks, millions and millions of grassy heads, all sprouting the first shoots of the season: cotton grass, *Eriophorum angustifolium*. Peter and Howie observed the caribou switching their diet to these fresh plants. Peter began speculating on the relationship between *Eriophorum angustifolium* and the awe-inspiring congregation of caribou. He understood collective birthing to be a natural way to deal with predators, but he had been searching for reasoning behind the geographic and temporal choices the caribou made in calving on the Utukok. Biologists had varying theories on how many separate herds there were in the north, and caribou from Arctic herds were known to intermingle during winter months. At that time the calving grounds provided the only distinguishing marker for biologists to determine definitively which herd an individual caribou belonged to. Basically, if a caribou showed up on the Utukok in June, it belonged to the Western Arctic herd, not the Central Arctic, Porcupine, or Teshekpuk herds.

Peter was fascinated by the mother-infant relationships in migratory species, a theme that would become central to his later research. On the Utukok he was struck by the difference between migratory caribou and the migratory birds all around them. In 2020 he told me, "The Canada geese, the pintail ducks, and a myriad of other birds sharing the Utukok landscape had migrated there over far greater distances, but they had to stay there to build nests, lay and incubate eggs, and rear fledglings until they were old enough to join their parents on the journey southward again. In contrast, caribou mothers gave birth among the *Eriophorum* tussocks, but within an hour the infant was on its feet and suckling and, with its mother's urging, moving off with her and thousands of caribou on their migration. And their bond formed strong enough to keep them together among thousands of others. What a miracle of adaptation!"

Austin Thomas working near the sod igloo at Cape Thompson, fall 1961 (Howard and Erna Kantner Collection)

Eriophorum angustifolium also became important in Peter's research as he studied how areas of early snowmelt in the Utukok Uplands caused this plant to sprout early—just in time to provide needed nutrients for pregnant female caribou directly following their arduous migration, preceding birthing, and after birthing, and even provided calves with their first solid food.

Howie trusted his friend's abilities—scientific and otherwise—but his own thoughts were far away, on the horizons, watching a golden eagle twirling in the sky over a newborn calf; or two grizzly bears on a ridge, one dark and the other shimmering blond in the sun, obviously mating, and afterward lying on the tundra near each other as if they were in love; and the white dots of caribou, scattered here and there, seemingly everywhere. Howie's eyes loved the windblown unobstructed stretches of Arctic landscape. For him it felt strangely like home.

Those months on the tundra had affected both men. Surrounded by wilderness and the most astounding array of animals they had ever dreamed of studying, their appreciation for this land and its creatures was metastasizing inside them in very different ways. What was growing in Howie was the same desire that had sent him as a seventeen-year-old fleeing as far from Toledo, Ohio, as he could go. Now he wanted to go still farther, to leave science's narrow, confining version of studying animals behind, to move closer to the animals themselves. Until he lived with them. Ate them. Wore them. Knew them. He didn't want to leave this place. Peter, too, loved the challenges of trying to live with the animals and in those wild places, but he had perhaps a more impossible dream: that he could do those things and also have a career as a biologist.

IN LATE AUGUST ONE EVENING, A SHARP CHILL FELL. THE RETURN OF TWILIGHT BROUGHT a deep eerie blue glow to the sky. The men stared up at the twinkle of a star, the first visible in nearly four months. Standing outside their tent in the dimness, Peter and Howie listened as white-fronted geese passed overhead, calling in the night as they flew south.

"I've been thinking—" Howie stopped and twisted at his big thumbnails.

Slowly, he told his friend what was on his mind. He didn't want to return to Fairbanks when the project ended. He had decided to no longer pursue a career in biology. He'd found an old abandoned sod igloo in March in his travels on the coast, near Chariot, the proposed blast site. His plan was to overwinter there alone, hunting for himself and his small dog team. He asked Peter if he would ship his dogs and sled back north from Kotzebue.

Peter listened to his friend in the twilight and agreed to do so.

But, as so often is the case with wild dreams, things turned out somewhat differently. When my dad arrived with his meager belongings at the entrance of the sod igloo at Cape Thompson, an elderly couple from Kivalina had already moved in. Prior to freezeup, they too had decided it might be a good place to spend the winter, in their case because of the ample supply of drift logs for firewood and good ice for drinking water. Their names were Austin and Mabel Thomas. The man had one leg and spoke a little English. Standing outside the igloo, he nodded and listened, and invited the young white stranger inside. There in the dimness, Austin introduced Howie to his wife, Mabel. Mabel had been raised by her father, an *aŋatkuq* (shaman) who had taught her the ways and the voices of the land, and how to hunt with her brothers, but no English. When she saw the visitor, she hurried to put food out and to heat water for tea.

The couple fed Howie boiled seal meat, dried whitefish, salmonberries, and seal oil. After hearing his story, Austin shrugged and pointed to an open space on the dirt floor and suggested Howie carry on with his plan. In the old Iñupiaq tradition of stunning generosity, he welcomed this stranger, who had appeared an hour ago at their door, to join them for the winter.

That afternoon my dad began his new life, daily hunting for the three of them and their dogs. Throughout the fall and winter, he learned old ways from Mabel, listened to old stories from Austin, and discovered his own new feelings for the natural world. In the coming years, Project Chariot wasn't ever canceled, but rather deferred as modernity kept moving northward. My dad did return south, briefly, but this land was in his veins, never to fully fade. He soon returned to the Arctic and three falls later built a sod igloo and settled at Paungaqtaugruk, where I was born.

As children, Kole and I often saw him shake his head as he warned us of dangers out on the tundra: thin ice, open water, charging moose, bears, and blizzards. In the same grave voice, he spoke of another danger out on the landscape of life—the trap of having a full-time job. He shook his head and recounted how lucky he had been to have recognized that a career in biology wasn't for him. To him it was a prison: studying animals from the cramped back seat of a Piper Super Cub peering out a tiny Plexiglas window to count creatures on an earth far below, and later being confined in an office cubicle all winter, sifting through data gathered during brief summer field trips, and always—week after week, year after year—dealing with the ubiquitous memos and meetings, procedures and politics.

More than any person I've ever known, my dad loves being outside, interacting with nature. There's something he gets from the natural world, some peace, or energy, or companionship. He can't stay inside for long without pulling on his footwear, grabbing gloves and a tool—an ax, shovel, machete, or rifle—and heading out.

Looking back across the years, I recall him in sealskin pants that Mabel sewed for him, and mukluks that Mamie made, and other furs that Mama and Elsie Douglas in Ambler sewed. He's always been sentimental, and many of his fondest memories were made in furs, footsteps that shaped his future, and mine too.

Along that journey, he has remained unfaltering in his curiosity, respect, and quiet compassion for his companions here, the animals and plants and other creatures sharing the trail—mosquitoes, bears, and barbed branches included. I used to think this was simply eccentricity; I wished he could be more of a fierce and focused hunter, what I was attempting to be. It was only later, after I incorporated a camera into my hunting and had to work exponentially harder for each photograph than I did for meat, and then started writing—and found gathering from the wilderness of words to be even more challenging—that I poured more thought into his intense appreciation for the natural world. Where did it originate? I decided that an absence where he was born and raised had caused it. Toledo, and the strict Catholic schools he'd attended, both were places where wilderness was more or less synonymous with danger and evil, where most ties to nature had been lost, where wolves and bears don't hunt beside humans, and the wind and the sky don't decide one's daily endeavors.

But, now I think that was too simplistic. Now when I'm alone in the wild for too long and talking to my surroundings—maybe congratulating a tree that grew up on this same hill with me, or asking a maggot her goals ("To fly, like your parents, right?")—I realize there is more. Now I think it was a need for connection that Howie was starving for in the city. Connection, and a sense of place, and of course truth. Starving for truth—that universal currency of the natural world, something humans never cease inventing ways to distort, cloak, and hide from.

For Howie, the pretend and the politics of people were never interesting. He was willing to pay whatever price to not have to spend his life boxed in by those things. When he found the tundra, a place where he was surrounded by truth, he found home.

CHAPTER 9

A NEW KIND OF CARIBOU

The immense herds of caribou here in the north began dwindling rapidly in the second half of the 1800s. Once plentiful, caribou were growing scarce. The increasing dearth of this species, vital to the way of life of the indigenous people—coupled with cultural upheaval and the increasing devastation of alcohol and smallpox and other diseases brought by white people—contributed to the destruction of ancient and fiercely won boundaries between Native nations. Tragically, too many people were suffering, starving, and moving to survive, for anyone to defend traditional borders.

As tribes and their territories began falling into disarray, larger world geopolitical borders were also in flux. Russian America (aka Alaska) was up for sale. It had been for almost thirty years, with no buyers. The Russian tsars realized they better sell the territory before Britain or the United States simply took it. Also, they faced the growing problem of riffraff: whalers, explorers, prospectors, and others were roving around Russian America as if they owned the place. Meanwhile, the young nation of the United States was embroiled in its own family feud, the Civil War. After the war ended, in 1867 the United States

A lone caribou left behind by a larger group rests on the main ice.

grudgingly coughed up a bit more than $7 million for the property, unknowingly making one of the most lucrative real estate purchases in history.

At 586,000 square miles, Alaska cost less than 2 cents an acre, and came with more coastline than the entire contiguous United States, as well as hidden riches beyond the ken of even that genie of old Aladdin's. The territory brought a new frontier to trample, grab, wreck, or maybe—it remains to be seen—treat with a bit more respect than that previous one of the Old West. What followed the passing of papers wasn't exactly jubilation: William H. Seward, the prescient senator who arranged the deal, got rewarded with nationwide ridicule; the Russians lost their foothold on this continent; and Natives in the north knew little or nothing of this newest transfer of "ownership" of a large hunk of the planet, in this case their home.

And the caribou? The caribou didn't know their lives were now further entangled with those two-legged creatures who had so recently appeared, or that both species were in for unimaginable changes. A faster-than-ever ride down the river of time.

DURING THOSE TUMULTUOUS DECADES, IÑUPIAT ESKIMOS HAD GONE FROM HUNTING FOR themselves to now also doing so for trade with whalers, explorers, military groups, mapping parties, revenue officers, and others, all of whom were hunting too. In the case of caribou, harvests were not limited to bulls or even to adult animals, and in addition to getting meat, Native hunters walked great distances on summer ventures with firearms to hunt newborn calves for their skins.

Calfskins—soft, warm, and velvety in June and July—were used to sew children's clothing and were requisite for adult lightweight underlayer clothing worn during the coldest conditions. This traditional inner "shirt" required four to seven calfskins for an adult person, and the outer parka worn over the shirt took three or more full-sized caribou skins harvested in August or early September, before the hair grows long and brittle. Because caribou hair sheds, breaks, and the skin wears through, these garments often needed to be replaced every year. And with each passing season, not only Natives but also more Outsiders were wearing caribou furs.

In addition to new weapons—exponentially more accurate, efficient, and deadly than traditional weapons—hunting practices were further altered by the degradation of tribal boundaries that had long limited caribou hunting areas for individual groups of Natives. Traditional Native hunting ideology, however, remained as it had been for centuries: when a resource was valuable and available, especially a migratory resource, it made the most sense to harvest as much as possible while possible.

Adult caribou were hunted for tongues, tallow, sinew, meat, and skins, and records from this period tell also of large numbers of calves harvested. An account by one of the earliest scientific explorers of interior Alaska, William Healey Dall, from June 14, 1867, observed more than a thousand bunches of fresh calfskins hanging to dry in a village on the lower Yukon River. Each bunch held four skins. Later, Charles Brower, the famous adventurer and trader who arrived in Point Barrow (Utqiagvik) in 1884 and basically spent the rest of his life there, also wrote about extensive use of calfskins, writing in July 1894 that residents of the community ceased whaling activities to travel to the Teshekpuk calving grounds to harvest skins. While in the fall of 1897 Brower wrote that caribou were plentiful, and he organized hunting parties that brought in an estimated 1,200 caribou carcasses for Outside whalers stranded there, in subsequent years he noted that caribou became surprisingly scarce.

Even as caribou populations plummeted, the land still held vast sources of food, as Minnie Gray and other local elders have attested; the coasts remained rich with sea mammals, the rivers full of fish. Birds and small animals also provided sustenance. But none of those creatures could provide what caribou had provided—that necessary combination of extremely light and warm fur, available in large quantities, and able to be replenished each year.

AT THIS TIME YET ANOTHER GROUP OF NEWCOMERS ARRIVED ALONG THE COASTS OF western and northern Alaska: missionaries. The timing of these newest Outsiders was opportune: Natives' lifestyles were literally in tatters, and local residents— dying and in desperate need—were very responsive to promises of a brighter and eternal future.

Spreading the gospel and an accompanying array of other good and bad cultural mores, many missionaries tried to improve the lives of the local people. They

P 1399 - Far Northern Eskimo Family, Alaska

An Iñupiaq family with their qayaqs (kayaks) along the beach in northern Alaska (John Urban Collection, Anchorage Museum, B1964.1.786). *Reindeer in a corral near Norton Sound* (Ickes Collection, Anchorage Museum, B1975.175.214). *A schooner and the US revenue cutter* Corwin, *modified into a merchant vessel in its later years, hemmed in by ice.* (Lomen Bros. photograph, Fred Henton Collection, Anchorage Museum, B1965.18.615)

brought medicine, built schools and hospitals, and worked to help feed and clothe a newly impoverished population. A well-known Episcopalian minister of the time, Sheldon Jackson, was one of the first to recognize the plight of the Eskimos and put the pieces together: their traditional reliance on caribou; the recent scarcity of those animals; and the suffering, starvation, and emigration taking place in the wake of the crash of the caribou herds. He came up with a plan to help.

In 1892, with permission from the US government, Jackson began importing *Rangifer tarandus*: in this case, reindeer. Unbeknownst to Jackson, they were the same species as the local caribou, only with the deer showing evolutionary changes from centuries of domestication and eons of geographic separation from Alaskan *Rangifer tarandus*. Ironically, Jackson and others brought these domestic deer from Chukotka, across the Bering Strait, that same area the first humans crossed to arrive on this continent, the same water the Russian fur traders later crossed, the same strait American whalers sailed north up to the Arctic.

For decades after, reindeer and reindeer herders became the cattle and cowboys of the north, playing a vital part in feeding and clothing the disrupted Native population and also shielding the wild caribou herds from further pressure by various groups of humans.

Reindeer thrived on the open tundra of the Seward Peninsula and other coastal communities, surviving grizzly bears and mosquitoes, wolves and hard winters, and providing what Jackson had envisioned: food, furs, and even some of those strange confining things the Outsiders cloaked their lives in—and Howard Kantner warned his sons about—jobs. Jobs, another part of Sheldon Jackson and other missionaries' strategy to assimilate Natives into Western ways.

As a kid, I heard tales of reindeer herds told by hunters who spent evenings around our fire: stories of reindeer corrals, roundups, and even sleds pulled by teams of deer instead of dogs. Like the stories about the absence of caribou, these accounts seemed impossible, outlandish, and when it came to harnessed reindeer, a bit too close to Santa Claus and Rudolph to take seriously. Still though, Clarence Wood and other Native hunters from upriver and downriver spoke covetously of reindeer; they claimed the meat was fatter, more tender, and better flavored and that local hunters were always on the lookout for stray reindeer mixed in with wild herds of caribou. It wasn't until later, when I spent time on the coast and saw old gray corrals and met elders who had been herders, that I discovered that these tales of the old days, like others, were packed full of history and truth.

AS THAT CENTURY OF UPHEAVAL NEARED AN END, THIS LAND WAS AWASH IN CHANGE—WITH newcomers, new technology and new "caribou," God, guns, starvation, and disease. What the Alaska Territory needed most was a breather, a reprieve to sort out these too-rapid changes.

Instead, in 1897 news was sledded south by dog team and then by steamship on to the States: Eight whaling ships were trapped in the ice at Point Barrow! The whalers were forced to overwinter, and the Natives had no choice but to hunt to try to feed them. The US Army stepped up to help, starting the Overland Relief Expedition and commissioning a reindeer drive from Wales to Barrow to provide needed food. Missionaries and herders along Alaska's western coast acquiesced to donate livestock and assistance. Afterward, while still in the glow of hard-won success and human cooperation, more startling news spread north, and soon circled the entire globe: Gold had been struck on the Klondike and along the shores at Nome.

Overnight, thousands of gold-crazed Outsiders boarded boats with prows pointed north. Behind them, a hundred thousand more fortune seekers flocked from around the world—more concentrated, needy, greedy, ill-prepared, thinly dressed, and disorganized than any of the previous groups of humans in history to arrive here. For better or worse, Alaska was now on the map, and a new century was on its way.

In shockingly short order, mere months, gold seekers had made their way up the Bering Strait—including hundreds of men on wood-burning schooners who traveled up the Kobuk River, past Paungaqtaugruk and Onion Portage, disrupting the Native culture with a barrage of new materials, metals, tools, technologies, rules, religion, and even some employment. These Outsiders found little or no gold, and most quickly vanished to the south.

The discovery of gold in the beach sand south of Wales at Nome caused a sudden city of tents to appear, and within a year Nome (aka Anvil City) was the largest community in Alaska with twenty thousand residents as well as bars, wooden boardwalks, hotels, and soon even electric lights. The surrounding area, the Seward Peninsula, long a barren and windblown tundra home to caribou, was now populated by an unprecedented number of humans and by growing herds of those recently imported animals, reindeer.

Gold brought a tidal wave of technology north: steamships sailed the coastlines during the ice-free summers, stern-wheelers plied the Yukon River, and behemoth gold dredges were freighted in to gnaw at the land, swallowing earth, ingesting tiny morsels of gold, and expelling thousands of tons of tailings. Crews stretched telegraph wires from the lower States—through Canada to the Gold Rush town of Fairbanks and then on to Nome—establishing the first blips of an electronic link with the Outside. This development greatly surpassed even the previous ultramodern communication provided by supply ships in the ice-free season and dog-team mail service in winter along the Yukon and north to Kotzebue, Barrow, and other communities.

The arrival of so many people and ships demanded, and made possible, a strange new phenomenon in the Arctic: the widespread reliance on food imported from faraway seaports. (Our modern-day issue of food security, a twenty-first-century catchphrase that in actuality means "food insecurity," took hold and continued to increase year by year in Alaska, until today where—regardless of all the hunting and fishing Alaskans do—it is estimated that only a fraction, 5 percent or less, of our annual food requirements are now harvested in state.)

BY 1910 MANY GOLD SEEKERS HAD GONE HOME. THE HUMAN POPULATION IN AND AROUND Nome and other nearby communities had declined drastically. The population of reindeer, however, continued to grow. The reindeer experiment appeared to be working. The animals were thriving, providing food and clothing and even jobs for locals, and conditions for northern Natives appeared to finally be improving— until 1918, when supply ships transported the worldwide influenza pandemic to coastal Alaska.

When the ice formed and snow fell, the deadly germs were further spread inland by dogsled. The flu grew worse, and all travel was banned. Armed guards were posted outside villages—including here along the Kobuk—with orders to shoot anyone who tried to pass. In the end the quarantine largely failed, and waves of sickness, starvation, and death again enveloped the north. Native camps and small communities suffered some of the highest mortality rates in the world, with some villages entirely wiped out, and an estimated 8 percent of the indigenous population in northern and western Alaska left dead in the wake of the pandemic.

Nome and surrounding communities had barely recovered when they were struck again in 1925, this time by diphtheria. The news sent a chill through the traumatized population of Alaska. Word of the epidemic flashed to Fairbanks over telegraph and was quickly carried to small villages and camps. Dog mushers in separate locations came up with a plan, to set out with teams of dogs to try to stop the outbreak by transporting diphtheria antitoxin serum by a sled relay across hundreds of miles of wilderness. Through supreme effort, they transported the medicine from Nenana, where it was delivered by the new Alaska Railroad, on to Nome in five days. The men and their dogs became heroes in newspapers in the lower states. Balto, the lead sled dog on the final stretch, was soon the most famous canine in history, with statues in both New York City's Central Park and downtown Anchorage.

Through these tsunamis of human turmoil, reindeer continued to thrive. Actually, the animals had become big business, second only to commercial fishing in the territory, and their numbers rose to a heyday in the late 1920s when nearly 1 million domestic reindeer grazed the tundra between the Kuskokwim River to the south and Barrow to the north, with the majority on the Seward Peninsula. Incidentally, this number matched the estimated number of caribou in 1800 and surpassed recent highs in 2000, when nearly 750,000 wild caribou populated the northern part of the state.

The Lomen brothers in Nome owned the largest herd of reindeer in Alaska, and the Canadian government noticed. Short of wild caribou, recognizing how reindeer in Alaska were flourishing, and wanting to help feed starving Inuit in the Northwest Territories, the Canadians signed a contract with Carl Lomen, the "Reindeer King," to drive three thousand reindeer to the Mackenzie River delta on the Arctic coast. Herders set out from near Kotzebue in winter expecting the drive to take eighteen months, traveling on foot in summer and by dog team in winter, to move the enormous herd across the top of the Alaska Territory. The herders spent the first summer here near Paungaqtaugruk, awaiting freezeup for better travel. By early winter the drive encountered the first of a series of devastating hardships (aka everyday life for caribou) that prolonged the journey. When they did finally arrive—after five years—more than half the animals had died or been lost, and replaced by calves born along the way, and far to the south the crash of the stock market had caused the price of reindeer to plummet.

During these same years, increasing numbers of predators were decimating Alaskan reindeer herds. Reindeer didn't migrate, were easier to catch than caribou, and their fawns were born a month earlier in the spring—perfect meals for ravenous grizzly bears coming out of their dens in April and May. Wolves also aligned well with reindeer; wolf packs found they could set up housekeeping year-round in one area, and hunt and den and raise pups all near the predictable stores of food. As a result, wolf populations exploded. Caribou were rebounding, too—likely because of reduced pressure by humans and other predators—and by the 1930s were beginning to spread back onto their former range.

Ironically it was these wild herds that spelled the final doom for the reindeer era. As caribou returned to their traditional territories, they mingled with the newcomers, and when they migrated on—as caribou always do—private herds of reindeer vanished with their wild cousins. Herd owners—businessmen, church groups, and Natives—fought both nature and the faltering American economy. The federal government stepped up in yet another attempt to assist Natives in the north, passing an act in 1937 making reindeer ownership legal only for Alaska Natives. By then many of the non-Native herders were finding reindeer no longer to be a lucrative business, were losing money, and now had no choice but to bow out of the industry. A second act established a bounty on wolves, but the struggling reindeer industry never fully recovered.

Caribou were finally returning, and across the land in the small communities, camps, and settlements of northern Alaska—like dots in this vast wilderness, far from the hubbub and availability of manufactured goods, and separated by poverty, great distances, and tough-traveling terrain—Native people held tightly to the shreds of their traditions.

CHAPTER 10

CARIBOU COMMUNITY

After the ice is thick enough, my family prepares to travel twenty-five miles upriver to Ambler for Thanksgiving potlatch. We talk about the trip for a month, and prepare for a week previous, including filling out order blanks to Sears and Roebuck and writing letters to distant relatives and friends. Mid-November often marks the first severe drop in temperature, and my dad worries about our family being stranded in the village. In the past, we traveled by dog team and had to camp the first night at Onion Portage, but now like most families in the villages, we have a snow-traveler. Ours is a Bolens, a red metal beast with a single-cylinder engine. All the snow-travelers are exotic and cantankerous contraptions, but ours is the strangest of all, with tracks in the front and skis in the back. It moves only slightly faster than our dog team did, and doesn't need to be fed caribou along the trail, but has its own thirsts and complications.

On the morning we leave, Mama packs caribou pemmican, extra sweaters, socks, and mittens. She heats the Thermos, fills it with boiling water, and forces it down into a cozy tube she's sewn from caribou skin. She helps Kole and me dress in our bulky gear and quickly pushes us outside before we get overheated.

After crossing a long stretch of ice, caribou pick up speed as they see and smell the safety of the tundra.

Bundled in our heavy parkas, overpants, and soft-bottom mukluks, we race back and forth like tiny spacemen. Our cold-weather mukluks are sewn hair-in; inside we wear caribou socks with grass packed underneath, and our feet have a buttery bounce when we run on the snow, light and soft and warm.

Howie hitches his twelve-foot dogsled to the snow-traveler and ices the sled runners with a basin of warm water and a piece of caribou hide. This old sled is the one he traded two wolverine skins for with Minnie Gray's husband, Arthur. While he works, the Bolens idles loudly, warming up. Heavy gray exhaust billows under the tracks as if the machine has caught fire. Finally, my parents holler for us to climb into the sled.

Kole and I run and clamber in headfirst, our feet hanging over the top rail. Howie whips and pats and brushes at our mukluks, getting every grain of snow off the seams where the moose-hide bottoms are sewn to the caribou leggings. Snow could melt and soak into the skin, causing us to get cold, and also damaging the fur. He wraps us in a sleeping bag, on caribou hides, and cinches a tarp tight around us. In the sled, Kole sits in front of Mama, and I sit in front of him, terribly confined and unable to move or see sideways from inside our bulky hoods, hats, scarves, and wolverine ruffs. We squirm, excited beyond belief to be going to potlatch, but at the same time nervous to be heading toward the creatures we understand least: people.

Kole and I know that in the village everything about our family is different and therefore wrong. We're white; there's no way to deny that. Also we talk wrong, own too few store-bought items, and wear too many furs. Our caribou mittens that Mama sewed to protect us from the extreme cold sum it all up: crude, utilitarian, and unattractive, the skin is gray and grubby, with the stitches showing—and because of past terrors along the trail that brought on tears and tiny frozen fingers, she's sewn our mittens to our sleeves. We can't remove them—and worse, she made them without thumbs!

The confinement in the basket sled is absolute, and we boys paw at the tarp with our mitten hooves, trying to press the tarp down enough to see the sky, the snow, or a glimpse of the trail ahead. The tension is huge—at any moment wind, weather, a broken runner on the sled, or a mechanical problem with the Bolens could cancel this whole trip. Canceling would mean Thanksgiving spent at home, surrounded by our familiar sod walls, caribou-skin cushions, wood furniture, and the worn words of conversation in a family of four living alone on the tundra. Canceling would mean no trip to the village until after the sun comes back.

IN AMBLER AS OUR SLED COMES TO A STOP, I BLINK, STUNNED AND STARING THROUGH frosty lashes. I must have fallen asleep along the trail. We're here! My hood is cinched tight, my wolverine ruff frosted, my scarf hard and icy with frozen breath. Behind me Kole squirms. We struggle to peer around. We're beside Tommy and Elsie Douglas's gray plank cabin. I see dark trees against the sky, and cabins up on the ridge with smoke rising from stovepipes. Dogs are barking everywhere. I hear shouts and running footsteps.

A crowd of boys and girls arrives at our sled, gripping the top rails, laughing and teasing and shoving. I recognize a few faces—Sammy Woods, Martin Cleveland, and a tall boy named Clarence Cleveland. I'm uncertain of the other kids' names. Quickly I lower my eyes in shame because of my white face, bright red cheeks, and fur clothing. Howie unties the sled ropes and flips open the tarp. Kids pelt him with questions: "Where you fellas come from?" "Aay! What kinda snow-traveler?" "Is that Kole?" "Is that Sef?"

A snow-traveler roars up beside our sled. The engine dies and the driver jumps off, brushes snow off his arms, and greets my parents. No one shakes hands. He asks about the trail. Kole and I stand nervous under the scrutiny of staring faces. Howie helps us down onto the snow. More kids run down the hill, and they all race back and forth, playing a big game they call "tag." Occasionally they pause to surround me. There are too many faces now for me to recognize any of them. They are all Eskimo, and wearing jeans and light shoes and tattered jackets; they point at my mukluks and parka and jeer at my thumbless mittens, reminding me what I already know: our family is different—and here different is bad.

The air is cold on my cheeks, the temperature dropping fast with night coming. Fish racks line the shore, and caribou carcasses are piled on logs beside the Douglases' house. Nearby a huge old husky is chained to a dead spruce tree. The dog has brown stains under his eyes, and he circles on his chain, whining and returning to

Ambler village after freezeup in the mid-1960s as winter settles in (Photo by Don Williams)

crouch protectively over a shapeless hunk of caribou with grooves gouged into the rock-hard meat from his gnawing teeth.

Tommy peers out his doorway. He ducks back inside for a jacket, and his wife's face appears in the window. Elsie waves, beckoning us in. Kole and I pad toward her. Bundled and helpless in our mittens, we keep our gazes down. The snow under our feet is hard packed, trampled with the tracks of boots, sled runners, and snow-traveler cleats. The barking of dogs echoes off the buildings, and in the chilled air I hear the crack of an ax and see a figure in front of a cabin, chopping. The sound is different than chopping wood—hard whacks on what sounds like stone, probably bone and frozen moose or caribou meat. Farther up the hill, someone is yanking repeatedly on a chainsaw that refuses to start. From behind Mark and Olive Cleveland's cabin, in the cottonwood trees, comes the sudden rumble of Charlie Jones's big white snow-traveler, the biggest in the village. All around the village, sled dogs howl in chorus with the sound of his engine.

Elsie pulls us inside and helps remove our fur clothing. Howie is fond of Elsie—he often refers to her as his trapping partner because he enjoys being able to bring her furs for her sewing—and now from the sled he unpacks a wolverine and otter skins he's brought for her.

Inside, Tommy and Elsie's house is crowded. Clothes are heaped in piles, boots are scattered on the floor, and Blazo cans and boxes and buckets clutter the corners. Below the window is a wooden table heaped with store-bought items. The window is iced over—nine small panes separated by wooden slats—all opaque with frost. The woodstove is beside the door. Kole and I stand near it after Elsie pulls our parkas over our heads. Our faces are stinging and red, and we cup our cheeks with our palms. There's no place to hang our furs, and we glance around, worried our scarves won't dry. Mama shakes frost off them and piles our clothes on the floor. On the shelves bright packaging catches our gazes: 7UP, Carnation, Tang, Swiss Miss, cherry Jell-O, Planters mixed nuts, Hills Bros. coffee, Clorox. The air is sharp with confined smells: the odor of untanned skins, soap, sour meat, smoke, dried fish, rancid seal oil, sourdough, bedding, and a chamber pot behind a cloth curtain. Kole and I stand stunned, from the cold and confinement on the trail, and now from the smells and strangeness of being in someone else's home. It's been long months since we've entered a human structure besides our own.

The Douglas boys, Peter and James, come in and Elsie serves coffee. Howie and Mama sit on a bench. They still have their sweaters and overpants and mukluks on, and they cup the warm mugs in their hands. Kole and I move close and lean on their thighs, shy and nervous around people. Elsie stirs hot chocolate and hands us hot mugs. We grip our cups and blow on the liquid to cool it, staring into the sweet, luscious steam. Tommy has a wide smile and large ears and moves agilely, refilling my parents' coffee. Kole and I watch, not quite believing what we've been told: Tommy has a wooden leg. We wonder, *Which one is it?*

At the table Elsie cuts caribou backbones with her *ulu*. She scoops water out of a big metal barrel, into the simmering soup. Into her palm she measures salt. She follows this with quick pinches of dried yellow noodles, and then slides the chopped meat and backbones off the cutting board into the pot.

Tommy and Elsie ask about the trail, and the adults discuss freezeup and fall fishing and hunting of caribou and other animals. Our hosts are pleased to have company. In the villages, and out on the land too, people greatly value the company

down at our hands, instantly embarrassed because we're too late to murmur that strange word that everyone seems to know to utter at the appropriate moment.

Men step forward and grasp pots of food. People are talking and laughing now. Conversations around us are mostly in Iñupiaq, with occasional English. The men move down the rows of benches and ladle out portions. Caribou soup comes first. It has cooled and a layer of congealed fat floats on the surface. A second man hands out big round white pilot crackers. One each. Next comes trays of baked whitefish, and a bowl of chunks of *uilyak* (frozen, fermented, raw) trout, followed by bowls of dark red *quaq* (raw and frozen) caribou. Next is fresh-boiled ptarmigan, and then a man tilts a coffee can, offering spoonfuls of seal oil to go with the ptarmigan and the *quaq*. People raise their eyebrows to indicate "yes" or wrinkle their nose "no" to decline a serving.

Next comes a platter with slices of bowhead *muktuk* (whale skin with blubber), cut into tiny black-and-white strips. Expressions of pleasure and surprise follow the server. Kole and I crane our necks to see. Black *muktuk* is traded from northern coast villages and is scarce along the Kobuk River. The man serving it nods to my parents, and we recognize Alex Sheldon's jovial smile.

More food keeps coming: more meat, Jell-O, pilot crackers, fruit cocktail, baked sheefish, and *tiktaaliq* (burbot) liver. Mama murmurs softly to Howie, mentioning how the amount of white-man food has increased with each passing potlatch. As she's talking, a man offers scoops of beautiful purple *akutuq* (Eskimo ice cream) made with rendered whipped caribou fat and blueberries, flaked whitefish, sugar, seal oil, and a splash of cold water. Whispers move along the benches. From the next bench over, Minnie Gray explains to Mama that the soup coming next is reindeer. Minnie is perennially kind and thoughtful, helpful and welcoming, and Kole and I hold our bowls out, excited. Quickly, we taste it. We squint, and shrug, not convinced. We know hunters strive to shoot a reindeer if they spot one. We've heard hunters say reindeer have shorter legs, darker hair, or sometimes even white patches like a cow, and that the meat is fatter, tastier, and more tender than caribou. We know that here in the village shooting a reindeer is a status symbol—but this meat tastes like caribou to us.

A bowl of slushy orange salmonberries catches our attention, and orange canned peaches, and then pale peeled frozen raw grayling. And my favorite—slices of boiled caribou tongues and bumpy white caribou cheeks and lips. It's hard to keep all the food from mixing on our plates. Most of it is cold now, greasy and oily, sweet, and some with strong flavors. Not all the flavors are good together. Kole and I are especially displeased about the rancid seal oil tainting our portions of fruit cocktail. Fruit cocktail is magical. The epitome of the Outside world and store-bought food. The miniature portions are brightly colored, flawless, sweet, and perfect, and Kole and I marvel and wonder: *Was this once real fruit?*

Around us the men talk and brag, and tell hunting stories. Women balance plates and containers on their laps, feeding kids and trying not to spill. People nearby on the benches converse with my parents in stilted English, questioning them about the winter trail, the movement of wolves, wolverine, and if caribou are still passing. Every mention of animals has to do with location, movement of the creatures, fur quality, fat content, direction the tracks were heading, how many, running or traveling slow? There's no tourism in this culture, only a pure utilitarian view of nature; the land is an endless grocery store, and everyone here has known times when mile after mile, every shelf was bare.

Thanksgiving Potlatch is the most cheerful gathering of the year—after a busy fall of gathering—but like fall, it is soon over. The children are restless, racing around and growing loud, and as quickly as it began, people round up their families and bowls of food, zip their jackets, pull on their gloves and hats, and file past the crackling barrel stove, out onto the packed snow.

Outside, they exchange a last "Happy Thanksgiving!" and chat briefly, acknowledge the cold—"*Alappaa!*" (It is cold!)—and hurry off, disappearing into their small cabins to continue the shared feast.

DOWN AT THE DOUGLASES', MY MOM HELPS ELSIE PUT OUT FOOD ON HER SCARRED AND greasy plastic floral-print tablecloth. The warm soups have cooled, and chunks of *quaq* are thawing. The house is full of smells. Kole and I watch the remaining Jell-O as we share a bowl of lukewarm caribou soup. The fat on the soup is beginning to stiffen, and when we drink cold water, it instantly congeals on our lips and the roofs of our mouths. Elsie retrieves *uilyak* sheefish from outside to add to the

meal. At the table she rapidly cuts chunks with a saw and then her *ulu*. The little dices of frozen meat turn gray with condensation. She sets out small saucers of seal oil to dip the fish in.

The adults discuss the trail again, the weather, and the ice, and the conversation leads as it always does to stories of luck and hardships on the land. Our parents' and the Douglases' backgrounds are completely different, the language they use different, too. Regardless, they laugh a lot, respect each other, and enjoy visiting. I sit listening, eyes down, acutely aware that my parents are white people, from Outside, and Tommy and Elsie are Eskimos from here. I'm not either of those things, and I question often, *What kind of color will I be when I grow up?*

When the meal is over, Mama gets out the caribou calfskin—the one the wolf chased on the ice—and asks Elsie for advice on sewing a parka for me. Elsie dries her hands, leans forward over the skin, bending at the hips to grasp it in her strong fingers. Howie tells Tommy of the black wolf. Elsie pauses and turns to listen. She gasps, "Ahh!" and scolds Howie. "*Arii!*" (Oh no!) "Next time try get it!"

Howie smiles down at his hands. Inside, the cabin is dusky. Elsie glances up, nods at Tommy. "You should put light," she says. Tommy blinks and smiles, shoulders into his jacket, and disappears out the door. Suddenly the house is bright yellow. Up near the ceiling, two hanging bulbs have become pear-shaped suns. The lights blink, fade, and then brighten again. Tommy is out there fiddling with his old diesel generator; it provides electricity for a bulb or two in each of the half dozen cabins he has strung wires to. Other villagers, beyond the reach of his wires, navigate their nights the way we do at home—with kerosene lamps or Coleman lanterns.

The electric light is shocking and wonderful, especially with winter Darkness draping a heavy black weight over our days. Now the windows appear painted black, and Kole and I stare at our reflections and this strange phenomenon—how the bright bulbs make the world outside seem so much darker.

Mama gets out brown parcel paper. She unfolds it across the caribou skin. Tommy comes back in, hangs his gloves and parka behind the stove. Elsie leafs through bundles of old patterns: wolf mitten patterns on newspaper, mukluk

Fox tracks wind across the sea ice.

patterns on parcel paper, *ugruk* (bearded seal) hard-bottom patterns traced on cardboard, caribou socks, and two different beaver hat patterns, and more. Finally, she finds the one she wants and carefully unfolds it. The brown paper is soft and old, like thin worn leather. She searches her shelves for a pencil. Together she and Mama trace the child's parka pattern to the new paper. Elsie speaks in halting English. Her pencil pauses each time she glances up, searching for a word in English to explain how to sew the sleeves, shoulders, and hood. This calfskin will work, Elsie explains, although traditionally late-summer adult caribou skins were used for parkas. She laughs and says she now prefers sheepskin, sheared mouton, because it doesn't wear out and shed.

I swing my gaze to Howie, hoping he'll tell of living at Cape Thompson and the calfskin shirt Mabel Thomas made for him. Disappointingly, he remains silent and listening. He still has the sealskin pants he'd sewn himself there, but sadly the calfskin shirt is gone. I don't know where it vanished to; I have never seen one, never met anyone besides him who has even heard of such a garment. Howie described it as downy soft, reddish in color, lightweight, and very warm. Mabel hadn't sewn one in years and told him, even back in 1961, that people very rarely wore them anymore. She had made it to show Howie, but also to remind herself of the old days when this undergarment was common and of vital importance in Eskimo country, before wool and cotton clothing were introduced.

Tommy is seated at the table and speaks in Iñupiaq to Elsie. Switching to English, he explains to us how hunters used to walk north in summer, great journeys overland on foot into the mountains of the Brooks Range, to hunt caribou. Tommy's face is serious, his voice slow and mesmerizing, almost religious in tone. It's hard to tell if he went along on these hunts. The women were left behind to fish, he says, while the men traveled north in search of caribou. He tells of travel routes up the Ambler River, the Redstone, the Reed, the Mauneluk, and other rivers flowing from the north.

Mama glances up from their work. "Did they bring the meat back?" she asks. "How did they carry it?"

Tommy smiles and explains that he has walked far, hunted everywhere, but that those longer hunts to retrieve summer caribou took place when he was young. He tells of traveling that country by dog team and later by airplane where a crash killed the pilot and left Tommy shattered and broken, trapped until his foot burned off in the wreckage and the pilot's son was able to yank his body free. His voice fills with emotion and reverence, and he says that God spared him. Elsie asks him a question in Iñupiaq. Tommy nods, his eyes soft and moving slowly now. He tells of hunting *siksriks* (ground squirrels) for parkas and describes the *siksrik*'s larger cousin, the beautiful *siksrikpuk* (marmot), and the rocky terrain where marmots live. This leads him to his real passion: hunting for gold. Tommy has long been infected with lust for the white man's treasures—gold and jade and other minerals—and now he stands and moves to a shelf. He rummages in a small tin box, hands Kole and me each a rock. Tommy explains that we're holding galena and bornite. The rocks are surprisingly heavy, square, and roughly an inch on a side, and Kole and I peer at them, awed and silent, yet confused because we've only heard of these words as geographic locations: Bornite on the Kobuk, Galena on the Yukon.

I hold my rock, waiting for Tommy to return to tales of hunting caribou. The rock is obviously unusual, and very rare, but I want to learn about animals. I want to be a great hunter and travel north of the mountains. I haven't thought about why I want to be a hunter—it is simply something I know. Hunting will make me good, and accepted, and normal, and Eskimo. Tommy is pleased that we boys are staring at him, listening intently, and now instead of talking of animals, he tells of his many searches for the white man's gold.

BY MIDWINTER, MAMA IS NEARLY DONE SEWING MY PARKA. I'VE TRIED IT ON COUNTLESS times as she fits the torso and sleeves. Each time I pull the tunnel of fur over my head, I'm instantly warm in the press of thousands of caribou hairs. When I pull it back off, broken caribou hairs remain in my hair, on my shirt, and falling in the air to the floor.

Mama needs more caribou skin to finish, and Kole and I help her scrape hides with her *ichuun* (scraper), an Eskimo-style flensing tool fashioned from a short section of metal tubing sharpened on the inside, with a wooden handle carved to fit Mama's hand. When scraping, we place a second caribou hide beneath the skin

Erna Kantner heats water to wash clothes with Kole nearby. (Howard and Erna Kantner Collection)

we're working on, to soften the pressure against the floor. We scrape for hours to get the outer layer of fascia and any dried meat or fat off the hide. The process is similar for wolverine, beaver, otter, fox, mink, and other furs. Furs such as lynx and wolf are relatively easy to scrape and tan because the skin is neither fragile nor paper-thin like fox or rabbit, or thick and hard like caribou legging skins. Leggings dry stiff and curled, with edges everywhere that can tear.

Our progress is slow—on our knees, with our forearms working hard—but it's satisfying, and the shavings steadily pile up on the floorboards. Shrews run out from the dark corners to gnaw on the scraps. When the scraping gets too tough, we move to a new area or come at it from a different angle, or hand over the *ichuun* to take a break.

After we finish, Mama spreads sourdough batter on the surface of the hide. When she's done, she folds the hide skin-to-skin to allow the sourdough to soak in. She folds the hide in half again and puts a heavy chunk of wood on it. The scraping is over until tomorrow. She has much other mending and knitting and cooking to do, and the acidic batter needs time to soften the tissues of the skin.

The next afternoon we unfold the caribou hide and stretch it flat. It's sticky, not nearly as pretty as when we finished the first scraping. Slowly it begins to dry, messy and looking like what it is—a white crusty pancake-batter-smeared hide. By evening the batter is dry and we carefully scrape the entire skin again. Some areas are too soft, still wet, and we have to wait. Other areas are dried, and we scrape gently so as not to tear the skin. When we do punch a hole, it's a sudden shock—discouraging—but we can't give up and instead use the puncture to gauge how carefully we must proceed. The hole can be sewn up later.

The final step involves using only hands. Kole and I can't do this—our hands are too small—and Howie takes over, grabbing two fistfuls of the skin and rotating one fist forcefully and rapidly, like winding a crank. When that area softens, he grabs two more handfuls—again and again and again, moving through the entire hide. Steadily, the skin turns whiter and softer. When he's finished, finally Mama traces the pieces she needs for the hood of my parka. The hide is beautiful. It's been hunted and skinned by our family, cared for with respect, tanned by our hands and hard work, and now it radiates an extra warmth as if some nameless force flows from the animal to us, magnetic, irresistible, drawing our hands to want to stroke the skin, to touch the soft fur, or lay on it, or sew it into clothing to wear against our own skin.

OUT ON THE TUNDRA CHECKING TRAPS WITH HOWIE, MY PARKA IS LIGHT AND WARM, BUT cold wind still leaks in around my throat. Back at home, Mama cuts more tanned caribou skin with her sharp little *ulu*. She peels strands off dried bull caribou sinew, wets her fingers, pinching and pulling outward, making the sinew smooth and thin enough to thread in her needle. Her skin needle has a triangle shaft, made to cut through leather, and she wears a thimble and keeps a small pair of pliers in her lap to pull the needle through when it binds. She shears the caribou hair to half length with a pair of scissors and sews small strips into the hood. She's not having an easy time with the area around the throat and laments Elsie not being here for guidance. This fix will have to do, though. The days are short and cold, and we won't travel to the village until the sun comes back.

It's common knowledge that the hood and throat are not easy to get right, and that the best Eskimo skin-sewers make perfect hoods, where virtually no wind can trickle in around a hunter's face. This is important. The most successful hunters travel out in the most dangerous conditions to bring home meat and furs and firewood, and correspondingly the most respected women are talented skin-sewers whose stitches are the tiniest and tightest—windproof and waterproof—and whose fur garments are warm, beautiful, and precisely fitted.

Times are changing fast, but the Native women still possess ancient skills that border on magic, the magic to take wild animal hides—greasy, wet, and raw—and transform them into soft supple caribou-legging mukluks, wolverine-head mittens, or sunshine ruffs fashioned from tiny individual ribbons of wolf, wolverine, and otter skin. Elsie, Mamie, Mabel, and other Native women hold in their weathered hands a disappearing tie between the creatures of this land and the survival of their people. With each passing season, more stitches in that connection come loose, and each open seam allows more of that past to fade, like warmth into a cold wind.

In our small dimly lit sod home, Mama daily teaches Kole and me from books, about the other world. She sews while we study. Howie works with hand tools, sharpening his axes, repairing snowshoes, and building sleds. At his workbench, he files a pair of Sears and Roebuck pliers into a tool for Mama to crimp *ugruk* skin for hard-bottom mukluks. Occasionally he catches a fox, mink, otter, wolverine, or lynx, and he skins the animals on the floor below the window. He scrapes and dries the pelts, and he sews fur clothing too.

In our heads Kole and I are mixing the realities of the two very different cultures, often without realizing that vital portions of each are missing. The Outside world remains distant, magical yet confounding, and only minimally experienced— where somehow hunting is less important than a haircut, young people are idolized instead of elders, and death apparently is not an everyday part of life.

In the nearby Native communities, the two cultures are colliding. The fault lines from that impact are leaving fractured ground for children, and the adults who might guide those children are themselves in the midst of this earthquake shaking their society to pieces. In this colossal upheaval and loss of cultural values, one imperative from the past whispers, echoing and reverberating: *Hunt!* It absorbs other fading imperatives, and continues to glimmer through the confusion until it becomes the embodiment of the old Native culture. It morphs even further, sheds original purposes and meanings as traditional connections between life and the land are lost, until a foggy chasm stands between this monolith and the ancient now-distorted forces that led humans to hunt in the first place: *hunger and need.*

CHAPTER 11

PERSPECTIVE FROM A GRAVEYARD

The night is cold and frosty, and again I'm standing out at the edge of a vast tundra, alone and listening to the immense silence, absorbing the arrival of winter and the Darkness. I shiver and shift, and the snow squeaks under my boots. On a tripod beside me, my camera perches, heavy and black and frozen. I'm on a high ridge along the Noatak River where I've built a small cabin for my daughter. A few yards away in the dark are crosses, grave markers, my only companions for the last month.

Overhead, a faint green glow condenses like fog in front of the stars, and I consider attempting a photo but I don't move toward my camera. From down the hill at the shore comes the eerie sound of new ice sheets settling, booming and echoing as huge fractures spear across the frozen river. Instinctively I glance over my shoulder in the darkness at the wooden crosses, rounded and thick with frost. I search in myself for fear and find none. I never knew these elders buried in this frozen ground, but they feel benevolent.

My gaze roves farther, peering about for a late grizzly bear who has been out recently. It feels like an acquaintance of sorts, too, although one I trust less.

Frost coats wooden grave markers along the Noatak River.

I turn back to the open tundra. The aurora has brightened. The outlines of mountains are visible now, reaching tall against the night sky. This is a big place, a broad valley where hills and mountains ring the vast plain of the Noatak flats. Fog is forming down in the flats, and I peer into the night, straining to spot the dots of caribou. There is no movement. I see only dark lines of brush and the black silhouettes of small spruce trees scattered here and there, standing in snow and cold silence.

I imagine caribou out in this night—with no warm cabin nearby, no woodstove, no light, no AM radio. I picture a small herd: some animals alert, others resting, sleeping a few fitful moments and waking with a start to stare wide-eyed into this darkness—this night and every night—forever on guard, aware that each instant is eligible for a surprise attack, to be chased by jaws and claws, torn to shreds, killed, eaten.

What sort of life would that be? I wonder.

To the north, a single faint light is visible, and gone, and visible again—a beacon on the airstrip at the village of Noatak, twenty miles away—a piercing reminder of the modern world. A Noatak man, an acquaintance named David Kelsey, told me this ridge once held a reindeer herders' camp. I don't know that history, but tonight, standing here, I can feel the old days, and the future too.

The blinking light has faded, hidden by fog. My mind travels farther north along the curve of our planet. Can it be true that the massive industrial Prudhoe Bay oil fields are this instant humming on the Arctic coast? And farther away, the megacities of the world—are they too out there tonight, LA and Detroit, Lagos, London and São Paulo, all happening right now? Are there truly billions of people, cars, houses, computer screens, iPhones, airplanes, cement sidewalks, neon signs, smog, gum wrappers, and tossed beer bottles? The answer is yes, I know. Still, it seems impossible here in this dark and frozen silence. I haven't seen a human in a month, and hardly any animals. The grizzly bear tore through the cabin wall before I arrived, made a mess of my tools and belongings. Since then I've seen it, or maybe its cousin, twice in the distance—brown on gray days, down along the shore in wet white snow, digging up desiccated salmon on the sandbar, sniffing the wind, deciding—or not deciding yet—to den.

The fall storms have been intense. The last one tore off my stovepipe, and for thirty-seven hours I thought it would take the roof. Since then the calm has been overwhelming, silent beyond a silence most humans will ever know. The land is freezing, finally—a month late this year—and facing winter wearing a million moth wings of ice crystals. Day after day the spruce below the ridge stand motionless, heavily bundled in parkas of frost. Each evening the temperature plummets, the sun sinks lower, and I know that soon I will lose my bright golden friend behind the mountains.

I haven't seen a caribou the entire time I've been here. Not even a track. Tonight, again I can't help fantasizing a herd will appear. I have an old battered .243 rifle, but hunting is not on my mind. I'm after something harder to acquire than meat, but I can't define it. For a quarter of a century I've been a wildlife photographer, and a writer, too, driven by a fearless and unwavering wolverine perseverance to show the world the importance of these animals and this wild land. And still that doesn't explain why I'm here, tonight—unless those efforts have not been enough.

Given the chance, I'd happily wolf down fresh meat, and I long for a photograph silhouetting caribou under the northern lights. But those are old, well-used desires. My camera is off, my gun leaning inside the cabin, and I'm looking into darkness for something different.

FOLLOWING PAGES: *Sandhill cranes fly low over caribou trapped on shifting ice.*

PART III

THE THIRD SEASON OF THE YEAR

On an afternoon in early February, a change is in the air. The sky is pale and cold, the tundra still white with drifted snow, but above the line of timber along the far riverbank—there—a yellow sun!

In the past weeks the sun has grown stronger each day, and now like a toddler learning to walk, it is floating free of the snow-covered land, bright and happy. Before long it sinks again and drops behind dark blue mountains. The sky glows and slowly fades to bluish-green twilight. Night returns, cold and black and glittering. By dawn a north wind is roaring, the temperature is minus twenty, and a ground blizzard shrouds the tundra. The land is furry gray, indistinct, bitterly cold, and the sun rises behind ice crystals, a cold orange face. Soon it descends into the frenzy. Above the storm the sky streaks pastel orange and gold. In the morning, clouds have moved in, and the following day is overcast and snowing, and the next—until a morning when the wind has died and dawn comes early, clear and chilly. On the tundra, twigs are fat fingers of frost. New snowdrifts squeak underfoot. The sun rises surprisingly fast, returning to flood golden light on the land.

By late March the sun is glaring bright, too intense to stare at, and the land is infused with light. Winter has lost its Darkness but holds on to its chill, and mornings are frosty, or windy with searing cold. In the long afternoons, the sun works to faintly warm the air. Along the shores, ptarmigan perch in willows like white lumps of snow against the blue sky, chuckling: *buck-buck, buck-a-buck.* Their feathers are puffed against the cold; their white furred feet grip branches as their sharp black beaks peck hard willow buds. Below, on the snow, snowshoe hares sit as still as porcelain statues, sunbathing in subzero air. The area around them is dotted with turds and pocked with tracks—theirs, and the bouncy tracks of mink and marten, the winding, careful footsteps of foxes and lynx, and the straighter trails where wolves and wolverine have followed the shores—each creature traversing the varied distances of their home territories, hungry and hunting for food and, in many cases that other need, a mate.

On the river, moose move deliberately, dark on the white ice, large and harried and haggard, fearing the trap of deep snow, yet needing to enter the willows to feed. Behind small islands they gather in herds, stomping down "yards" where they can maneuver and venture cautiously into the thickets to bend and break and swallow the endless branches needed to stay alive.

April brings more snow, and the land shines with a fresh new layer of paint, and the white tundra stretches to white mountains. Each day is longer and brighter, and traveling is the best it will be during the entire year. The land is huge, endless, and inviting after the heavy confining tarpaulin of winter storms, cold, and the Darkness. In all directions the world beckons, a sparkling new planet with no boundaries, no borders, and no clock other than the shining sun. And like the roaming wolves and the migrating caribou, humans, too, inhale all this sunlight and cold air, and a yearning rises in our blood, to roam, to travel beyond our home horizons.

Night is nearly gone now, growing short and thin, losing to the huge, invading day, and in the sky the familiar stars say goodbye one by one. In leaving they offer little warning—unlike that first star to return in August—of the unimaginable changes coming with Arctic summer.

And sure enough, one night comes a forgotten and fearful sound, the patter of rain. By morning the snow is wet and settling fast. The alders and willows sway in a strange new way—thawed, no longer frozen solid—and the trunks of spruce are dark with wet bark. Birch branches glisten and drip that long-absent silver substance, water. Grizzly bears will be coming out of their dens, and any day now the spring birds will begin to arrive. It is time for the caribou to come north. The herds could cover fifty miles a morning on this snow. Already patches of wind-blown tundra are melting out, and brown heads of tussocks mottle the landscape. Soon all will turn to mush and slush and melting ice.

Finally, a line of dots appears on the white tundra. A female leads a dozen cows and calves, her small hard antlers curved and brown. As they move closer, the animals' flanks are pale and sun bleached and knobby with warbles growing large under their skin. The group moves steadily north, disappearing over a ridge. But the feeling they have brought remains. They are so good to see, so welcomed, these strangers who feel like friends. With the return of the caribou comes a feeling of safety and providence.

CLEAR WEATHER BRINGS COOL NIGHTS, CRISP MORNINGS, AND SUN-DRENCHED DAYS. IN the late evening the sun rests for a few hours and a chill fills the air, refreezing the snow. Bears stroll on the surface crust, moose plunge through into grave danger, and in the night groups of caribou pass quickly and leave only curved prints in the hard snow and occasional turds scattered like black seeds on the white tundra. Wolf trails crisscross over the caribou trails, and as the night crusts harden, their tracks show less and less, until morning when they reveal little of where the wolves came from or where they have gone.

In the warmth of the day, a large herd appears on the white river, while in the distance downriver another smaller herd has halted on the ice. Both groups are soon gone, traveling fast, as the snow softens, and pale blue-green streams of over-flow spread on the river ice. At the mouths of sloughs, dark brown water shows and meltwater fills the snow. A few days later the center ice floats, as moats form along the shores. In the thickets rabbits chase mates, and a tiny unfamiliar bird flits in the alders—not a winter resident, a chickadee or a redpoll, but a first spring visitor—a male junco. Overhead, a bit of cotton floats in the air. But no, that's a moth! And there, a mosquito. Oh, insects—that other forgotten force of nature!

Lines of migrating caribou cross the tundra now, their trails becoming trenches in the snow. At the south bank of the river, the trails end at water. The animals crowd and mill by a dark moat, awaiting a leader. The cows stare across the current and a few step forward, stare north sniffing the water, and turn to weave their way slowly back into the herd. A lone female moves to the shore, repeats the process, and she too loses her burst of courage. Caribou begin lying down, resting, waiting, showing no interest in leadership. Suddenly panic sweeps the herd and a gray wave of caribou sprints along the shore. As quickly, the stampede stops. Caribou mill and run back and forth. No animal appears to know who started this panic, what first caused it, or whether they should continue fleeing, and in which direction. Maybe it was a squirrel leaping to a limb, or a bent willow melting free, or an eagle flying overhead. Finally, the group stands still, the animals appearing thoughtful, almost dejected as one by one they again drop to the snow to rest, and await a leader.

Geese and ducks fly over the ice, calling, searching for melting-out ponds and exposed sandbars, and often swinging low over the caribou as if drawn to their

In spring, waterfowl flock to Silver Dollar, a tundra lake that has become a grass field as a result of drainage caused by melting permafrost.

fellow travelers. Sandhill cranes broadcast pterodactyl commentary from the distant tundra, and high in the blue sky—invisible at first—drifts down the faint cackling conversations of snow geese, winging their way north. Robins shriek and rustle in the leaves on the hillsides, sparrows and warblers sing in the treetops, and marsh hawks and falcons and eagles soar overhead. Everywhere in the sky, hurrying small shapes fill the bright night air, and only occasionally does a raven cross the river, dark wings panting, the familiar black silhouette flying straight, moving with purpose, working, busy raising an early batch of babies. The gray jays—local residents too—have also already nested and are strangely absent in the rush of so many tourist birds dotting the sky and landscape.

Daily, the ice grows more dangerous, dark in areas and collapsing into floating tinkling musical icicles. Icebergs surge up from the depths—frozen to the riverbed all winter—suddenly loose and roaring as they surface and brown water cascades away. The current slowly takes hold, turning them gently, steering the ice pans on their new journeys.

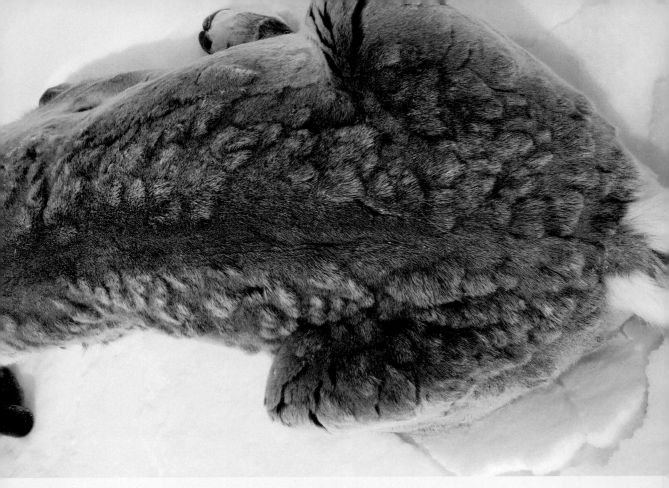

By May, warbles appear as large knobs under a caribou's hair.

Near the resting herd, in the low evening sun, a cow with one antler stands and walks down to sniff the water. She wades decisively in. It hardly seems possible, after so much inefficiency, fear, wasted energy, and lack of leadership, but animals begin rising to their feet. Suddenly hooves are splashing and a river of caribou flows into the river of water. The herd swims toward the center ice where the snow has settled and melted and left a broad firm surface that stretches miles upstream and down. Already the leader is struggling to get her forelegs up. Bucking and heaving, she uses her chest and forelegs to break loose ice and clamber up. She shakes, shakes again, and doesn't look back as she trots across the white ice sheet toward the far shore.

THE WORLD OF SNOW AND ICE DAY AFTER DAY DETERIORATES. CARIBOU HALT IN THE MIDDLE of the river in large and small groups, detained but undeterred. Randomly, individuals plunge through the ice and struggle to climb back out, while a yard away others stand and calmly stare at their dangerous surroundings. Here and there caribou fold their front legs and lower themselves to the ice, as if in the midst of all this danger, resting, or maybe meditating, makes more sense than moving. Gulls land beside the animals, peck up fallen warbles and turds, and rest after their own long migration north. The sun reflects glaring light off the melting snow. The air is warm, and the water rises steadily.

One afternoon the entire sheet of ice begins sliding away, hundreds of yards wide and stretching out of sight upstream and downstream, a solid landscape of ice familiar for the last eight months and now abruptly and physically leaving. In bends, and in shallow areas, the ice remains jammed, while in other areas open water stretches shore to shore and the huge blue current sparkles, otherworldly after winter. Suddenly summer is no longer a forgotten season. Long lines of caribou take the opportunity to swim instead of crossing ice. Soon more ice appears from around the bend. A soft roar fills the air, millions of icicles and thousands of pans jostling, bumping, pushing, grinding together. Trees along the shoreline shake and struggle. Willows are dozed under as sheets are forced over the banks. Incrementally, the ice slows. The roar and tinkling of needles diminishes. Finally, all movement stops. A white-crowned sparrow sings from the top of a spruce. On the far shore caribou continue to file out of the willows. They stand and look across a quarter mile of terrifyingly jumbled ice.

All afternoon the jam holds, creaking and tinkling, while more caribou flow from the willows and crowd the snowbanks along the shore. The jam raises the water, and sloughs and lakes and low tundra areas flood, reflecting the night sun like molten metal. A north breeze blows a chill across the land. Feathers of new ice form on the dark water. The sky blazes fire behind the mountains. Near the caribou herd, a moose appears on the crisp snowdrifts, not sinking in, relaxed and reaching high up to pluck sweet pussy willows. Farther along the shore, a porcupine clings in the branches doing the same, and not far from it, a sow grizzly with three tiny cubs stands and claws down armloads of willows to munch the sweet buds.

Snipes roll and dive overhead in the night sky, relentlessly announcing their newly claimed territories, and from the flooding ponds comes the cackling of hundreds of white-fronted and Canada and snow geese, holding huge raucous midnight parties, hollering and fighting and mating.

Winter is nearly over. Breakup is at its peak, big and breathless, one season giving birth to the next, and nothing can hold back the water now. And nothing, it seems, will hold back the caribou. Through the night, tendrils of animals wind their way across the rotten ice, long lines of caribou following caribou following caribou. And suddenly the entire river of ice is moving again. The caribou stop. They stand and await fate. The ice picks up speed and carries them down the rumbling river, around the bend, traveling on toward life or death.

Ice flows all through the night and the following day. Again, caribou crowd the south bank. In the morning the river is wide gray water, speckled with small white pans. The herds have vanished, crossing in the night. The migration is nearly over.

For the next week or so, small groups occasionally appear, swimming fast across the wide water—mostly bulls now bringing up the rear—with dark velvet antlers showing, growing fast. The land is changing fast, too; poplar trees are putting out sticky buds, and fireweed and bluebells have sprouted from the warming soil, and the wild roses, willows, birches, and alders show tiny tips of coming leaves. The water drops daily. The current lessens, and the river is quiet. Along the shore, mountains of beached ice randomly cascade perfect clean white icicles into black muck.

In the heat of the day, the birches and blueberry bushes begin to sparkle with green diamonds. In the night the first frogs in the ponds rattle, freshly melted out and broadcasting joy or desire or hunger, and in the new stillness loons and grebes call, both telltale sounds that winter has melted and gone. Spring, too, is almost over, and it's sad to let go of those seasons. Summer lies ahead: a transformed land; a land strange with the sounds of liquid water, wind in leaves, birds singing, and bugs humming; sweet scents in the air; flowers; and so much sun. And the quiet intensity of so many lives with much to accomplish.

Caribou cross solid ice and plunge into a final moat before they reach the north shore.

A group of caribou finds its way across jumbled river ice. A willow ptarmigan on the tundra with spring plumage beginning to show

FOLLOWING PAGES: *A young grizzly sow brings her cubs out of the den for the first time. A female muskox protects her young calf. Raven chicks beg for food. A tundra swan and white-fronted geese pause to feed as they migrate north in spring.*

Ice pans are forced along by rising meltwater. A mink drags his prize, a muskrat, through slush and ice. Ice jams along the Kobuk River cause the water to rise and flood the nearby tundra.

CHAPTER 12

BIG CHANGES ON A BIG LAND

Yet another sudden shift in reality came to the Arctic in the 1920s in the form of a distant hum from the south. A dot appeared in the sky. In seconds the hum swelled to a roar and an airship thundered over the sod igloos and tarpaper shacks of Kotzebue. Hundreds of huskies howled in unison. Villagers ran outside, staring up in awe and consternation.

The ski-plane circled and touched down, bouncing across snow-drifts, stuttering to a stop. A crowd raced onto the ice and cautiously approached the strange craft. Men peered at the wooden prop and struts, the wings and engine. A schoolteacher and a preacher stepped forward, shook hands with the pilot, and asked for news of the Outside, and where the man had come from, how long was he staying?

By 1927 Ralph and Noel Wien had established a business in Nome serving Deering, Candle, Kotzebue, and Point Hope, and soon small planes became the magic carpets of Alaska, reaching every corner of the territory, connecting far-flung communities by new trails in the sky. Bush pilots' names and exploits became legendary. Here on the frontier, these men—and a few women—owned the sky and the sea ice,

Blizzards leave scalloped snow in the western Brooks Range.

distant mountain ranges, rivers and lakes, and whichever animals they chose to take; they could climb over the clouds, literally soaring over common folks who remained mere specks down below, stuck following old trails on foot, snowshoes, and plodding dogsleds. Animals were surpassed as the fastest creatures on the tundra. Now, like eagles, humans could traverse unimaginably harsh landscapes in mere hours. A person could climb into an aircraft in Fairbanks, then climb out onto the ice at Kobuk. Before long mail service by dog team was discontinued. Medicine, news, supplies, and even people could now be sledded *through the sky!* Sick and injured villagers could be transported to hospitals; miners and explorers could be dropped at random points in the wilderness; even *food* could be hauled north, high above the land.

Not long after airplanes arrived, bush pilots began ferrying north yet another batch of Outsiders. These elite strangers came from around the globe and were willing to pay top dollar to be dropped deep in the wilds, kill a desired animal, and as quickly have a plane pluck them back to civilization, usually with the large antlers, hide, or a once-snarling head of whatever they shot. This new tribe quickly became warmly despised by Natives for competing for animal resources. Before long these Outsiders had acquired the nickname "head hunters" because of the headless animals they left abandoned near their camps.

These days, with so much powerful technology at our fingertips, it's hard to imagine how worshipped and relied on these aircraft and their pilots were in the wilds of Alaska during the twentieth century. And even today, some of these same aircraft—fifty-, sixty-, and seventy-year-old Piper Super Cubs on skis, floats, and tundra tires—remain what they were back in the day: awe-inspiring flying sleds. This Model T Ford of aviation continues to be one of the best airplanes ever built for accessing remote areas of Alaska, and a vehicle of choice for transporting hunters, tourists, hikers, rafters, and photographers far into caribou country.

BY MID-CENTURY, TRANSPORTATION ON THE LAND AND WATER WAS ALSO EVOLVING. THE first outboard motors had come north; slow and heavy, they were infinitely faster than the previous options: paddling, or lining a skiff upstream with dogs and family members pulling ropes while fighting bugs, brush, cutbanks, and current. Evinrude, the first brand of outboard motor, was adopted quickly into Iñupiaq diction, as were the Primus stove, Thermos, and that other wonderful hissing smelly invention, the Coleman lantern.

The chainsaw arrived—louder and smellier still—an incredible luxury item, with its throaty roar ringing out along riverbanks where cabin dwellers cut firewood and house logs. And then, a hundred years after the introduction of rifles came an invention with an equal capacity to alter reality in the north: mechanical snow-travelers.

The arrival in the 1960s of widespread personal motorized transport—snow-travelers (aka snow machines, or snowmobiles) and the earlier outboard motors—distorted the time it took to hunt, the distance that hunters were able to travel, and other details of the traditional cultures' relationship to the land. Unlike aircraft, which remained out of reach for most villagers, snowmobiles were relatively simple and inexpensive, and nearly every local hunter could purchase and pilot one. Previously, rifles had increased hunters' ability to harvest animals while still retaining much of the focus on the land. Food and furs had remained essential to survival; transportation by sled dogs required more of the same: meat, fish, and fat. The new outboard motors, and especially snowmobiles, proved incredible tools for harvesting these needed resources, while at the same time diminishing the need for huge quantities of protein to power dog teams. Local hunter-gatherer societies, in embracing these machines, soon discovered that the machines had to be fed things that didn't grow on the tundra: gas, oil, and dollars. Also, the snowmobiles and outboards—many built as American recreational vehicles—didn't hold up under the harsh use by Native hunters in rugged Arctic conditions. Costly repairs requiring more dollars became a fact of life; gas and oil rose steadily in price, and new machines were continuously being manufactured: faster, sleeker, more reliable, and more desirable and expensive.

In the wake of this newest upheaval, hunting pressure increased on wolves, wolverine, lynx, foxes, and other furbearers, a long-trusted source of cash from the wild. Coincidentally, in the 1970s and early '80s the prices of furs on national and international markets rose, briefly helping offset the burgeoning local demand for

It was another quarter century until the influx of prospectors brought the first federal game wardens north. In small communities where life consisted of gathering from the land, folks were confounded by the concept of hunting regulations—as opposed to a person's abilities, need, and luck naturally limiting their harvests. They soon learned to fear game wardens like bad weather or an unexplainable disease, and dealt with the threat in a similar fashion. Lawmen had adopted airplanes for means of transportation, and local hunters learned to watch the sky. When game wardens couldn't be avoided, hunters used a technique they'd honed over thousands of years: they waited and resumed normal life when the "weather" improved.

In the wake of World War II, with the aid of aircraft, aerial hunting, trapping, and poisoning of wolves contributed to the growth of caribou herds. When the territory became the forty-ninth state in 1959, the federal government still owned most of the land, but enacting and enforcing hunting and fishing regulations across the state became the charge of the new Alaska Department of Fish and Game (ADFG) based in Juneau. With statehood, the industries of commercial fishing continued to grow, as did sportfishing and sporthunting, which ADFG was charged with managing. Subsistence hunting—largely outside the range of harvest reports and subject to shifting external forces—was harder to quantify.

Meanwhile, in northern Alaska, with each passing decade, geologists discovered more minerals—vast deposits of lead, zinc, copper, gold, coal, oil, and gas—under the tundra the caribou lived on. Developers remained stymied by the sheer remoteness of these riches—until, along the northern coast, test wells at Prudhoe Bay struck oil in quantities rivaling the Middle East reserves: tens of billions of barrels of crude oil. Development of those resources required a megaproject, the eight-hundred-mile Trans-Alaska Pipeline, which first required the settlement of aboriginal land claims between Natives, the federal government, and the state. In 1971 Congress passed and President Richard Nixon signed the Alaska Native Claims Settlement Act (ANCSA) creating thirteen Native Corporations, deeding Natives land and monetary compensation, and paving the way for the Pipeline and other large-scale extractive development.

Out here on the land, the lifting of a pen didn't instantly alter people's lives. Alaska was still a frontier. And private ownership of land was not a Native custom. Sudden, invisible changes in land status by distant unknown agencies was ethereal, and old news; it had happened before when Russia had basically "sold the world." Life had gone on, and did so now. No fences appeared on the vast landscape; folks continued hunting, trapping, and traveling, eating caribou, skinning seals, picking berries, and feeding their families from the land.

Biologists employed by the new state continued to study and attempt to quantify the huge and far-flung Arctic herds of caribou. The southern herds—especially the Fortymile and Nelchina herds whose range intersected the road system—were suffering from increased pressure by humans and as a result absorbed much of the money, attention, and management efforts of ADFG personnel. The large northern herds remained distant, something of an afterthought, and attempts to sort them out—as far as distinction between separate herds, range, and population numbers—had consistently proven difficult. This was before modern radio and satellite collars (which came into widespread use in the late 1970s and mid-'90s, respectively), and before modern photo censuses where individual caribou are manually counted from photographs by human eye (and recently even by computers), and biologists struggled on the ground and in small aircraft to count large numbers of animals spread across a territory the size of the moon, or the Louisiana Purchase, or something equally wild, remote, and ridiculously gigantic.

Early censuses were hard to conduct, and not always accurate. Throughout the 1960s the Western Arctic herd had been estimated to be approximately 100,000 to 200,000 animals, and it wasn't until the 1970s that two large Arctic herds, the Central Arctic herd and the Teshekpuk Lake herd, were definitively identified as separate herds from the Western Arctic herd. Human harvest at the time was an even tougher number to determine and was estimated from the 1950s through the early 1970s to be about 25,000 caribou annually for the entire Northwest Arctic, which referred to essentially the entire northern half of Alaska—say 300,000 square miles, an area larger than Texas, with some of the wildest terrain left on earth. While this harvest estimate seemed high, at the same time the herds appeared to be thriving, and in many northern game management units (GMUs), the state did

Dog mushers cross the inner sound near Kotzebue.

not set a closed season or proclaim a limit, and liberally allowed the taking of cows and calves. Some biologists speculated that the introduction of the snowmobile would lower caribou harvests because of the reduced need for food for dog teams, but that remained to be seen. For now, the hands-off approach by the state showed all the signs of working.

Farther south, rapid changes were taking place in the larger Alaskan towns and cities. The Trans-Alaska Pipeline, one of the biggest projects in human history, was under construction, and for much of the 1970s the state's politics and budgets—including funding for caribou research and censuses—were dictated by the frenzy of the new oil boom. In urban areas, the human population rose quickly, and with it pressure on nearby game. Both the Fortymile and Nelchina herds suffered population collapses in the 1960s and '70s, developments that biologists and game managers working for ADFG were strongly criticized for. What happened with those herds would in part affect what happened next with the northern herds:

ADFG employees were becoming gun-shy of politics, and the politics of caribou were about to get much worse.

In 1970 the estimate for the Western Arctic herd rose to 242,000. Then in 1976 the census dropped precipitously, with an initial minimum estimate of 64,000 caribou, later revised to 75,000. At the same time, aerial surveys in 1975 and 1976 near northern villages reported alarming quantities of dead caribou—nearly 1,000 wounded, dead, and abandoned each year. Biologists believed that the actual numbers were higher than those observed and decided that humans (aka Native hunters) and wolves were responsible for causing this sharp decline in caribou. The crash of the herd in the late 1800s and the half century it had taken to recover weighed on their minds, as did what had taken place with the Nelchina and Fortymile herds. Game managers and the state Board of Game moved quickly to try to remedy the situation and to avoid being blamed for the demise of another large and important herd.

I was eleven when my family and others heard the shocking news: Hunters were to be allowed only one caribou for the entire year! Previously, Howie had shot as many as eighty or one hundred each fall to feed us and his dog team, and for *paniqtuq* and pemmican to last the summer. Caribou were part of the landscape; nearly everyone lived off caribou, more or less. Now across the north, families faced this proclamation from ADFG—which, for people living off the land, was as frightening and ludicrous as if the law had suddenly limited each of us to one blueberry a year.

Coincidentally, that fall, caribou poured through Onion Portage. Villagers also reported another long string of thousands of caribou flowing down the coast past Point Hope, Kivalina, and on down to the tent community at Sisualik, passing Kobuk Lake camps and Kotzebue, moving south to Buckland and beyond. For many lifelong residents of Northwest Alaska, incongruously, this was the most caribou they had ever witnessed near their communities.

I remember how unsettling and scary it felt when the State of Alaska issued Howie and other hunters a locking metal band to be clipped onto the right hindquarter of their one lone caribou. And how tense that winter was for many families. I remember hunters carrying those silver metal bands, holding them in their big hands, showing them to each other—carefully not clicking them closed and instead saving the band for the next hunt, and the next. How else could one caribou feed a family for a year? Fish and Game issued only three thousand or so silver bands for the entire north of Alaska and later noted "widespread noncompliance and blatant violation of the regulations." But from our local viewpoint, we had to eat. And hunting now meant being extra vigilant: listening for airplanes, working fast, burying gut piles, and kicking snow over blood trails while watching the sky with fear of being arrested.

Rumors and fear swirled through the villages. Was it true that it was illegal to leave a skinny animal? Was it true that it was illegal to feed caribou meat to sled dogs? Did that include scraps? How about guts? Could you get arrested if your dog was gnawing a frozen hoof? What about a sickly animal—could a hunter leave it, or feed it to his dogs and go shoot a healthy one for his family? (Incidentally, yes, wildlife managers, in an attempt to further protect the herd, had put forth a proposal to make it unlawful to feed caribou meat to dogs, which inadvertently—or otherwise—severed yet another link to Natives' former lifestyle.)

Finally, in the summer of 1978, fair weather and that other variable, money, aligned to make an aerial survey of the Western Arctic herd possible. The count came in at 106,000 animals. Good news—except, unfortunately for some people this increase in just two years seemed too good to be true, almost a biological stretcher, and it was widely believed by residents of the northern part of the state that ADFG had missed a group or groups of animals in the previous census. (Statistically, caribou don't have twins, and populations fluctuate depending on quantity and quality of food, cow mortality, calf survival, disease, weather, predation, and other factors. For the Western Arctic population to remain stable, several figures, more or less guidelines, are considered necessary: a calf survival rate [referred to as recruitment]of 15 percent, a parturition [pregnancy] rate of 70 percent, and a survival rate of 85 percent for adult cows. Although herd growth is possible and not uncommon, growth at the rate recorded from 1976 to 1978 is somewhat less likely.)

All good intentions on behalf of the department were buried in the ensuing fray of finger-pointing, as was the fact that the caribou population had been declining, and—regardless of our human hardships—most likely did benefit from reduced hunting. Tight restrictions were eased, albeit slowly, and in the following years hunters were allowed two caribou, and then three, and eventually what we have today: five caribou a day, every day of the year. Basically, open season, with virtually no restrictions on hunting practices.

The state's initial heavy-handed attempts to remedy the situation, however, remained controversial, and have not been forgiven or forgotten. Criticism rained down on ADFG—probably harsher critiques than if the herd had crashed. The department was hammered, and still is, for cultural insensitivity, racism, endorsing starvation, attempting genocide, and more.

FAR AWAY IN THE BUREAUCRATIC WORLD, OTHER GOVERNMENT AGENCIES WERE BUSY with many more pressing issues that needed to be sorted out—especially if future

Retrieving vegetables and berries from the siġḷuaq *China and I built* (Photo by James Q Martin)

development was to proceed—and in 1980 Congress passed another act, the Alaska National Interest Lands Conservation Act (ANILCA), which established varying degrees of protection on more than 100 million acres of federal land. President Jimmy Carter signed the bill before leaving office, doubling the landholdings of the National Park Service (NPS)—43 million acres of Alaska were suddenly under its jurisdiction—and cementing Carter's position as this state's most hated man for decades to come. The wording of that act, especially ANILCA's accommodation of rural subsistence priority, was shaped and greatly altered by input and concerns expressed by residents of northern communities—voices arguably a little louder and more insistent in the wake of the caribou controversy. The effects of that act back then were complex, and even today it's difficult to live with or explain the political, cultural, and environmental quagmire ANILCA created. But in oversimplified terms: across Alaska people believed the federal government was taking what was rightfully theirs—the right to use and treat the land how they saw fit—and feared new laws would limit hunting and fishing, mining and development, and would destroy our various Alaskan lifestyles.

My family was among those people, and very afraid. None of us were absolutely wrong, or right. Probably no one person or group was quite as bad or good, virtuous or evil, as those on opposing sides saw them. But big government was now at the doorstep, and hunting and gathering, living off the land as a way of life, would never be simple again.

CHAPTER 13

A WINDBLOWN TRAIL TO THE FUTURE

Howie smoked a pipe when I was a toddler. He saved white ptarmigan wingtips, stuffed up in our log eaves, and he'd pluck out a stiff feather to clean his pipe stem, and then slowly pack the bowl with tobacco. Prince Albert, Edgeworth, or sweet-smelling Cherry Blend. Kole and I watched, barefoot and grubby, breathlessly awaiting the excitement of what came next. Staring down at his hands, he'd grin and murmur, "Boys? Give you a penny to light my pipe." And we scrambled for the matches he held.

I don't remember ever getting any pennies, but we loved striking those matches: the thrill of the spark, the smell of sulfur, and cautiously touching the flame to the tobacco. I already knew I wanted to be like him, a hunter and a trapper, with string and matches and ammunition in my pockets, old worn mukluks, and tools and guns leaning in the corners. At age five Kole announced he would be going to college. I said nope, no way would I be going. Already I was dreaming of a dog team and my first hunting rifle.

Howie taught us to set muskrat traps and skin ermine, peel logs, and build sleds and kayaks and the other things we needed to live off the

Caribou trails vein the snow in spring.

land. He taught us to hunt for food, haul firewood, and cut moss to insulate our sod home. It wasn't long until goose hunting was my favorite activity. The arrival of the first geese marked the beginning of spring—long days and bright sun—and it was exciting to snowshoe to the pond and build a blind out of grass and willows, and crouch inside, listening, holding our breath, watching, and calling to geese. When Howie shot, Kole and I raced to catch the fat tasty meals that came down from the sky. Beautiful, fresh exotic meals compared to old caribou and older fish that we'd tired of over the long winter.

He had a calm, pleasant way of teaching, and allowed us to learn on our own, often by accidentally cutting ourselves with our knives, or snapping our thumbs in traps, or getting cold or burned or hurt falling out of trees. We didn't mind taking off our boots and socks, and even our pants, to dog-paddle in frigid water and ice pans to retrieve geese—nor plucking and gutting the birds afterward. The burned-hair odor of singeing waterfowl became a sweet smell, tied to fun times, sunshine and tanned faces, and the best dinners of the season. As the spring snow melted we were often barefoot, running across patches of tundra, searching for bird nests, scavenging for old cranberries, nibbling cotton-grass shoots, and bending down willows to suck nectar out of pussy willows.

When I was nine, I built a birch sled. Kole and I got sled dogs, and I was allowed to use Howie's old Winchester .22. The rear sight was gone, and I made a new one by drilling a hole in the head of a stove bolt. But the gun never hit consistently, and finally I was allowed to look through the big catalog at Mark Cleveland's store in Ambler to pick out my own new .22. I chose a pump-action Remington Fieldmaster. Months later it arrived in the village in a long green box. Howie sewed a sealskin scabbard, Kole and I lashed it to our dogsled, and we explored relentlessly with our small team, trapping and hunting and learning ice.

After breakup, we put the sled under the cache and got out our homemade kayaks to hunt along the river and in melting-out ponds. Our friend Alvin Williams came down from the village, and together we portaged from lake to lake, through the bright nights, hunting ducks and geese, beaver and otter, gull and grebe eggs, and whatever else we could find. By then Alvin had a Rossi .22, and Kole and I had used a hacksaw blade to cut grooves in the old Winchester and affix a small, cheap blurry Gander Mountain scope to the gun. Suddenly, surprisingly, Kole had a gun more accurate than mine or Alvin's.

My parents laid down a few laws: we had to tell them where we were going, dress warm, and carry matches. We weren't allowed to kill anything that our family wouldn't eat or use—no shedding spring foxes, no skinny summer geese, no beaver after the fur was sun-shot and red—and we weren't allowed to wound an animal without going after it and doing our best to retrieve it. That meant no shooting caribou with my .22 rifle. I needed a bigger gun.

Years passed, Howie put his pipe down and never picked it up again, Kole and I trapped more and more furs, and still my dreams of a bolt-action hunting rifle kept getting delayed and detoured. Howie's philosophy was to live simply and wait. Waiting, in our family, was pretty much one of the Ten Commandments: Just wait. Best to wait. Let's wait and see. Also, any expenditure of real cash in our family was frowned upon, discussed, and then put on hold—to be discussed again later. Guns were expensive, and Howie had his Husqvarna .270, an extremely accurate rifle; he seldom misplaced an empty cartridge and carefully reloaded his brass; he was a good shot and experienced; it made sense to him that he shoot whatever caribou our family needed.

At that time, in the villages people were anxious to drop the old for the new, and many folks were seeking seasonal jobs to acquire cash. For myself, I only envisioned a future of hunting and trapping with a dog team. There were complications, I realized—thousands I couldn't yet imagine, and a few that I could—the largest being the flood of new laws marching north, affecting people who lived off the land. In my family's case that included the terrifying federal Bureau of Land Management (BLM), a government agency that had long threatened to burn our home—if it would even burn, being damp and built into and from the earth—because my parents hadn't claimed a homesite. Instead they had chosen to follow the traditional Native way, to simply live on the land and not have a paper title to a tiny piece of this earth.

Communication at that time between the villages and the Outside was limited, but news and rumors drifted north, of the Trans-Alaska Pipeline and proposed national parks that might soon blanket Alaska and make it illegal to live off the

Building my first sled at age nine; with a grizzly hide at age twenty-two (Howard and Erna Kantner Collection; Photo by Stacey Glaser)

land. The threat of restrictive regulations loomed large. But larger and closer still was the consistency of the seasons: the land and the animals, especially caribou.

Travelers and hunters regularly stopped in at our place along the river. Shyly, I stole glances at their guns, and occasionally asked about calibers, trajectories, and brands. In those years a rapid change was taking place in the rifles that local hunters used. People were switching to semiautomatic guns. The Ruger Mini-14 was the favorite and quickly became ubiquitous, carried by nearly every hunter, in every season, everywhere.

The Ruger was designed for law enforcement, military, and private security markets, and it was small and easy to carry, with a wooden stock and a removable clip that held a previously unheard-of number of bullets: twenty or thirty rounds, or even more. The ammo was essentially identical to that used by the US military. More importantly, the high-velocity .223 caliber was extremely versatile, and big enough that a hunter could head out with a Mini-14 prepared to kill whatever he encountered—goose, moose, caribou, bears, beaver, wolves, wolverine, and more—all with the same gun. In the case of larger animals, such as grizzly bears, killing the animal often required hitting it with multiple rounds. Most folks didn't see that as a deterrent; hunters weren't overly worried about wounding an animal as long as they got it, or there were more to shoot.

Locally, hunting "seasons" were based on when each species was fat and tasted good. State and federal bag limits and hunting regulations were not well understood and often became a concern only if a game warden was rumored to be in the vicinity. The concept of needing a paper license was baffling to many, and paying for the right to hunt seemed a moral injustice. Even the government's words for animals—"game" and "big game"—felt depraved. Hunting was no game, and locally the whole subject of licenses and regulations was viewed like thin ice: perilous and best to avoid. Outsiders, on the other hand, it was believed, should follow those rules. They were the ones who made them, after all. Fortunately, the land was large, wild, and not easy to navigate—and the rarest creature out here was a lawman.

A TROUBLESOME RIFLE CAME MY WAY ONE SPRING. I HAD HINTED AT WANTING, AND FINALLY was grudgingly permitted, to shoot an ancient gun that leaned in the corner, a dark brown well-beaten .30-06 Springfield from World War I. Howie kept the gun as a spare and occasionally carried it for bear protection when we walked the ridges or went berry picking.

The Springfield had a steel stock plate and iron peep sights. I was small and boney, not yet five feet tall, and the gun was long and heavy and awkward—it kicked hard. In Howie's cache in his wooden junk boxes, I found a carton of old military rounds. Rummaging more, I unearthed handfuls of sawdust, caribou hair, glass vials, and eight or nine corroded .30-06 reloads someone had given to him. The brass was mottled with dark corrosion, and the bullets were loose and had dropped down into the necks of the cartridges. Only the tips showed. I bit at the lead, yanked the bullets up with my teeth. But after chambering a round, I discovered that I had to keep the barrel pointed down to keep the bullets in place. It was difficult to keep this straight in my dyslexic brain because

A hunter on a snowmobile chases caribou on the tundra.

At that time, the 1970s, groups of caribou overwintered on the tundra south of Paungaqtaugruk. The caribou brought hunters from upstream and down. With the new snowmobiles, hunters could travel much faster and farther. Winters were exceptionally cold in those years, and my family stayed close to home in our narrow trails packed to our waterhole, woodpile, caches, and nearby trap sites. Howie trapped in much the same way that he hunted—walking on snowshoes and waiting for animals to come to him.

Over the next few days, the men were gone during daylight hours. They returned after dark to eat, rest, and recount their adventures. They came back jovial, their sleds heavy with caribou, their cheeks and noses frosted, their footwear crusted with ice and chalky red frozen blood. They told of driving many miles, pursuing caribou.

The white man, it turned out, was a construction worker named Dick, there to work on a new school being built in Kiana. He was amazed and intrigued that white people lived the way my family did, in a sod igloo, eating frozen raw fish and dried caribou, and wearing furs. He admired the wolverine Howie had trapped and the sleds he'd made by hand, and how Kole and I walked barefoot on the snow. He complimented Mama's cooking and thanked us repeatedly for our hospitality. "You don't know how much I'd love to live out here the way you do," he said. "You're so lucky. I'd spend the whole winter here, if I had the time. I just wish I had the time."

Sitting on log stumps, Kole and I squinted at each other, confused. The one and only thing that everyone we knew had no shortage of was *time*. How could it be that this friendly man—who had a splendid grub box with Jersey Creams and Fig Newtons (even though the cookies tasted slightly of gasoline) and fruit cocktail, and store-bought winter boots and a fancy snow-traveler that went faster than a caribou—have less time than other people? Howie felt bad for him, having to return to a job. We all did. Howie gave him a scraped fall caribou hide. Mama gave him a sack of *paniqtuq*. Kole and I carved wooden spoons for him out of spruce kindling. Still, after the men disappeared back downriver, we remained perplexed. We understood scarcity well, but had never encountered a person with a shortage of time.

THE DAYS GREW LONGER, AND CARIBOU BECAME HARDER TO FIND. HUNTERS TOLD OF abandoned carcasses on the tundra and a big heap of dead animals that had been piled up for bait. At our table, folks from upriver villages comfortably blamed downriver people. Downriver folks had the same to say of people from upriver villages. "They always waste," visitors confided. My dad cleared his throat, nodded politely. "More tea or coffee?" he offered. It was common knowledge that hunters were shooting pregnant females—in winter those animals were most likely to have back fat—and that it was also standard practice to leave a skinny animal.

Hunters told my parents the caribou now ran away at first sight of a snowmobile. "Chased too much," they said. It was the first we'd heard of such a term. We had known that hunters pursued caribou with machines, and chased down wolves and wolverine at high speed, but somehow we hadn't imagined quite the extent of what was taking place across the river on the open tundra. Our snowmobile was far too slow to chase anything, we rarely saw caribou in winter, and Howie wouldn't hunt them then. He preferred the flavor and the abundant fat found on

Howie dragged the cow to the sled. The caribou was skinny, a fraction of the weight of a huge fall bull. The meat was bloodshot and gritty with guts. Howie sloshed his fingers clean with snow and shook the bloody slush off his hands. He stared around the wide flat empty tundra. "C'mon, boys. Get your things. Let's go."

At home, we skinned the fetus and tacked the skin out to dry. The hair was downy soft and disappointingly thin, like mouse fur. The tiny leggings curled into brittle strips. Kole and I tried sewing with the skin, but it was papery and fragile and tore easily. Shrews rustled on the floor and crawled up the walls to gnaw our handiwork. We boiled some of the meat, too—the shoulders and hindquarters and the tiny tongue—but it all tasted bland. Blander even than red squirrel. We boys decided that chasing down animals with machines was not good, and that was not something we would ever do when we grew up.

THE WINTER BEFORE KOLE WENT AWAY TO COLLEGE, 1983, HE AND I TRAPPED TOGETHER for the last time. The population of rabbits and ptarmigan, moose, and caribou had been rising—partly because of increased vegetation and warmer winters—and correspondingly, furbearers were plentiful. Coincidentally, fur prices were also high. Kole already had a reputation as a quiet genius and soon would leave to study physics, mathematics, and that strange new fad, computers. Mama had spent as much time as Howie teaching us, and she had homeschooled both Kole and me nearly all twelve grades and worked to keep up with Kole's unquenchable thirst for learning. He was big and strong, mentally tough, eminently capable out on the land and equally talented on the vast landscape of knowledge from books. I was none of that—small, dyslexic, late to read, terrible at handwriting and organization skills, and only vaguely interested in school subjects. If pressed I'd admit interest in biology, or maybe being a veterinarian. Mostly I was good with my hands and wanted to be a trapper and a hunter. I had a few dozen traps, a homemade sled, four dogs, and a .22 rifle. Appallingly, I was sixteen and still didn't have my own bolt-action center-fire hunting rifle.

In February, after the sun came back and was yellow again, Kole and I mushed our five-dog team to Ambler to sell two wolverine skins. The trail was blown in, cold, and slow, and we had to spend the night. The next day we ran into a school-teacher who sold rifles out of his cabin. Kole and I were shy, uneasy around people, and socially inept, but the man coaxed us to come have a look.

The sleepy village had changed and was busier than in the past: a new runway had been built, and planes were coming and going. A court ruling had forced the state to build huge high schools in every village, and modern homes were being built for villagers. A well had been drilled, a pump house built, and most houses were provided with flush toilets, running water, and electricity. A city office had been built, and white satellite dishes erected, establishing communication to the Outside—first with one telephone for the entire town, and soon with telephones and TVs in most homes.

The teacher's cabin was past the creek at the upper end of town. It was one room and small and cluttered, with boots and white buckets lined up by the door—meaning the family hauled water the way ours did—but a TV was playing, and the place was bright with electric lights, and on the kitchen counter was a black telephone and a silver microwave oven.

The man informed me that he didn't have bolt-action rifles. My heart sank. Kole and I turned to go. "Hey, wait. Here, take a look at these." From behind the kitchen counter, the teacher thrust two guns at us. The rifles were black with plastic stocks, aluminum frames, and strange pistol grips. They looked exactly like weapons used for war, and Kole and I stepped back, uncertain if it was against the law for us to touch them.

"C'mon! Here you go!" The man insisted they were legal, and very reliable—the exact caliber as a Mini-14, only tougher and more accurate. "Let's step outside," he offered. "Fire a few rounds."

We pulled on our gloves and politely followed, though I knew I wouldn't ever buy a machine gun. The teacher fired single shots, rapidly, into nearby trees. I stared at the frozen spruce, feeling guilty and wordlessly apologizing for this rude awakening from their midwinter hibernation.

"Hey, watch this." The teacher put the butt of the stock to his forehead and squeezed the trigger. *Boom!* "See? See that? This rifle has very little recoil."

I stared, laughed nervously, and glanced down at the snow. Kole smiled and shrugged, mildly intrigued. He was going away to college and had no use for a rifle,

Alvin Williams ties a load in a basket sled.

wisps of snow moved on the tundra like smoke. We watched the small herd sprint away. "Athletes," Raymond commented. Alvin chuckled, white teeth in his darkly tanned face. I'd never heard that term and laughed, too—at how accurate it was, and because of how sacrilegious that sounded, in so many ways. The caribou obviously had been chased too much. Alvin moved with purpose now, unhitching the shackle connecting his basket sled to his snowmobile, nodding and pointing for Raymond and me to do the same. I fumbled with my crescent wrench, nervous. I'd never hunted in this manner. Never driven so fast that it was necessary to leave my basket sled. Never chased an animal with a snowmobile. I asked Alvin how far we might go. He grinned and shrugged, and yanked his starter rope. Raymond chambered a round in his Mini-14. Alvin did the same. They gunned their machines and roared off across the tundra.

I glanced around. The land was white, nondescript, and dotted with small shrubs. Low hills stretched away to the south. The Jade Mountains were miniature in the distance behind me. I couldn't believe how far my friends were willing to travel to hunt caribou. Quickly I started my engine, anxious to not be the one who came home empty-handed. I glanced back at my basket sled beside the other two sleds. I squeezed the throttle. Howie's snowmobile was narrow and tippy, and now went much faster than he'd ever driven it. Suddenly I was airborne. The skis slammed down, the handlebars twisted sideways, and the machine flipped. I flew off and landed face-first in the hard snow. My rifle banged my ear, and for a moment I was stunned. I jumped up, scooped the snow out from my wrists and neck, and righted the machine. Alvin had disappeared. Raymond topped the skyline and vanished. I started the snowgo again, now fearing for my teeth, my rifle, my machine. And my reputation as a hunter if I couldn't keep up with my friends.

Eventually, I spotted three caribou standing, watching me. The leader leaped up on its back legs—a caribou's warning of danger—and they raced away. I gunned the machine. Soon I was a hundred yards behind and felt a predator instinct take over, narrowing my focus and emptying my mind of all thoughts except catching these animals. *Go! Go!* Strangely, my feelings reminded me of my dogs—the way they were unable to stop once they were in pursuit of an animal.

CHAPTER 14

THE MOST POLITICAL ANIMAL

The northern herds continued to grow larger, expanding their range north to the Arctic coast, down to the Nulato Hills, and eventually even as far south as the Yukon River. Meanwhile, technology and bureaucracy were doing extremely well, too, expanding rapidly across the same territory as the caribou. With the passing of ANILCA—the largest single sudden expansion of protected land in US history—new national forests, parks, preserves, monuments, wildlife refuges, and other conservation areas were established on 157 million acres of Alaska. This was in addition to 104 million acres promised to the state with statehood, and 44 million acres deeded to the new Native corporations.

Immediately following the passage of ANILCA, in 1981 a new group of Outsiders appeared in the Arctic. Although they had sent a handful of scouts ahead to help frame the language of that congressional act, the majority of these newcomers stepped off passenger airplanes extremely green—with good intentions but no experience in local villages or out on this land, and no experience with local people and customs. They wore green uniforms and came with equipment, vigor, and righteousness in the form of a federal mandate to manage the gigantic

A hunter returns to Kotzebue hauling a sled loaded with caribou.

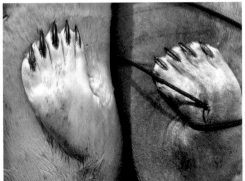

Hunters Andrew Greene and Randy Toshavik pursue ugruk *(bearded seal) on the spring ice. Ugruk flippers, one with a slit to secure a rope tie. Alvin and Clara's son Jeremy skins a bull caribou at Onion Portage.*

against the law—Howie certainly had never reported shooting it after it charged his wife and toddlers. Like villagers we were frightened of other faceless forces, too: Greenpeace, the Sierra Club, PETA—even the words "wilderness status"—all lumped together in local vernacular as "animal lovers," aka dangerous Outsiders who wanted to take away local people's food and furs, firewood, and way of life.

AFTER THE POLITICAL PARCELING OF LANDS IN THE WAKE OF ANCSA AND ANILCA, THE state remained in charge of overall management of fish and game until a provision buried in ANILCA promising "rural preference" for subsistence hunting and fishing on federal lands surfaced. In 1990 an Anchorage man sued, pointing out that the Constitution of the State of Alaska guaranteed equal access to hunting and fishing for *all* Alaskans. It was a touchy time, a touchy situation, with the state caught between its constitution and a federal mandate (specifically a US government treaty with Natives).

Federal agencies saw no choice but to step in and begin managing game laws on federal lands. Suddenly countless bureaucratic nets were overlapping, confounding all of us—rural subsistence hunters, urban sporthunters, state game wardens, NPS rangers, biologists, and land managers—with a maze of contradictory regulations.

Migrating caribou, and caribou hunters too, now unknowingly crossed in and out of the jurisdictions of state, NPS, BLM, US Fish and Wildlife Service, and other landowners including Native corporations, oil companies, private individuals, and any number of others, all of whom had varied rules and restrictions on who qualified to hunt in those areas. If this wasn't baffling enough, both the feds and the state claimed jurisdiction over navigable waterways, and in 1974 the federal government had passed another chunk of legislation—the Marine Mammal Protection Act—establishing a 100 percent closed season for non-Natives on polar bears, sea otters, seals, walrus, beluga whales, and other marine mammals, and a year-round limitless open season for Natives hunting those same species.

Seasons and bag limits—for caribou, moose, bears, sea otters, seals, rabbits, ptarmigan, ducks, geese, and even murre, gull, and grebe eggs—now varied depending on where you lived, where you hunted, *definitely* on who you asked, and, in the case of marine mammals, if you could claim to be at least 25 percent Native.

As misinformation swirled through isolated villages, cultural changes occurred on an emotional level as well. Intense generosity had been integral to Native culture, since time immemorial, so much so that feeling right and good about themselves required it, and was nearly as important to people as having enough food. But now, local residents' animosity rose sharply against regulations, scientists, agencies, Outsiders, and especially the lightning rod of things to hate: fly-in sporthunters, aka head hunters (now more often referred to as Cabela's Army). Natives questioned why non-Natives were allowed to hunt caribou when they weren't allowed to harvest a single seal—not even a dead one that floated up on the beach. Was the limit on federal lands fifteen caribou a day, and if so, why didn't that apply to Natives hunting on Native lands? And why was it legal for sporthunters to leave grizzly bear meat and only take the skull and skin? And in what kind of world could it be illegal to feed unwanted caribou scraps to your starving sled dogs?

Far away from life on the land, the bitter turf war was also growing worse between state and federal biologists and managers. Native political powers, wisely testing the wind, found more receptive and sympathetic ears with federal managers and soon shunned their state counterparts. Both sides had black eyes from messing with subsistence rights and were gun-shy of starting any additional conflicts with Native groups, and all agency personnel—regardless of their uniforms or lack thereof—upon arrival in northern communities discovered a disconcerting reality: here they were bad guys, all lumped together as "game wardens"; politics trumped truth, and questioning or busting or even bothering the wrong people—Natives—could blow up in their faces and, especially in the case of federal employees, could lead to a swift transfer south to guard the Liberty Bell or watch prairie grass grow in Kansas.

Caribou biologists had no choice but to recognize that their jobs were terribly entangled in politics. They left, or learned to walk a tightrope, where science often came second to nightmarish public relations in villages where they might be publicly shouted down and shamed for random unrelated past grievances. They did have better funding than in earlier years, and access to better technology, including radio and satellite tracking collars as well as more accurate photo-census capabilities. But to have any hope of making an impact locally and becoming accepted in the villages, they had to put in years—decades—and become part of the community in which they worked. Though often unrecognized for their efforts, biologists in the northern part of the state were actually extremely protective of subsistence hunting, a stance that didn't always sit well with their bosses in Juneau.

One example of success was here along the Kobuk, where the ADFG began collaring caribou in the river at Onion Portage. For decades the September migration remained predictable, and biologists were able to boat up to the animals, detain them briefly, attach collars, draw blood, and assess body condition, before quickly releasing the animals—usually cows because they provided much more information, and because the bulls' neck size changes so drastically over the course of the seasons. This project kept helicopters—reviled by locals—away from the migration, and peaceful relations were more or less maintained as hunters shot caribou in the river yards from working scientists. Joint state and federal agencies worked together too, and even included community outreach, bringing village high school students and teachers along for hands-on experience as they placed radio transmitters on the animals. While local elders remained largely mistrustful and morally opposed to collaring any wild animal, biologists, with these new devices, were able to increase their understanding of the seasonal movements of the growing Western Arctic herd and how that knowledge integrated with factors affecting all Arctic caribou populations, including natural predators, development, rapidly changing weather and vegetation, and increasing conflicts between human hunters.

With these new tools, computers, and still a hell of a lot of flying in small aircraft, ADFG could finally keep track of this massive herd, see where the animals wintered, where they were aggregating, and gather all sorts of additional unexpected data—including the fact that nearly every collar halted at the Red Dog mining road and turned back north for varying durations. Without collars it had been impossible to determine the extent of the range of the Western Arctic herd—a massive area stretching from nearly the Yukon to Barrow—or observe the correlation between range size and population. Biologists were also able to get accurate harvest reports from nonlocal hunters—mostly white men, who were accustomed to reporting if, when, and where they hunted and were thrilled to report a successful caribou hunt above the Arctic Circle. (In the case of the Western Arctic herd, the number of caribou taken by nonlocal hunters has remained relatively static and small, ranging from 374 to 799, total, over the last 20 years.) Still, though, a cloud obscured biologists' data, disrupting their numerical methodology: the number of caribou killed by resident hunters remained an unknown.

Native hunters—especially those out away from population hubs and more closely maintaining an actual subsistence lifestyle—had long ago learned that reporting harvests fell somewhere between demeaning, confusing, and sheer folly. For many, English was their second language. Life out on the land was already a daily balance of risks. And who wanted to compound that by being arrested for breaking unknown laws invented by pale strangers from a foreign land? For their part, policy makers far away in Juneau and Washington, DC, remained unable to comprehend the true cultural divide between Western society and traditional hunter-gatherer cultures. Sure, Natives were wearing Levi's. And yes, they adored drinking 7UP and watching *The Six Million Dollar Man* on television. But similarities thinned out from there. Most Natives continued to relate to the land in more or less the same way their ancestors had, where the only real professions had to do with hunting and gathering; the definition of success was bringing home fat meat, furs, and food from the land; and generosity and sharing resources were the foundations of community, social equality, and freedom from authority. Even the local culture's relationship with time wasn't like that of the Outsiders, with nearly all attention paid to the present, a fatalistic disinterest in the future, and a near-complete acceptance of death.

For most locals, hunting—*especially hunting*—was simply no business of Outsiders. Besides, reporting hunting to the agencies was dangerous. I learned that lesson as a seventeen-year-old. Unlike anyone I knew locally, I attempted to tag a grizzly bear hide. Alvin and Clarence and my other friends along the river generally skinned a bear and threw the hide in the willows to play it safe—they had little or no use for bear hides—and quickly shared the meat with family and friends. I instead foolishly wasted twenty-five dollars on a tag, which subsequently somehow turned out to be the wrong piece of pink plastic. When I tried to remedy that

Caribou migrating across the frozen ocean arrive at Kotzebue.

and acquire the correctly colored "big game tag," I was entrapped by a Fairbanks biologist who kept me in conversation until he could call enforcement officers. The lawmen arrived flushed and excited to charge me with a crime. They confiscated my bear hide and stood gleefully fondling the fur, commenting on how they'd never seen such a blonde. The following month, after the ice went out that spring, I was forced to buy a plane ticket south to court, where the judge threw the whole thing out, called it nonsense, and ordered the hide returned to me. My close call with the law taught me some valuable lessons, though: my friends and elders were wise, and I don't know how many bears I've shot since or whom I might have shared the meat and fat with, but I stopped reporting grizzly bears. Only thirty years later did I report one to ADFG, and only because a friend wanted to send the hide to a commercial tannery in Anchorage.

Mary Williams cuts meat for paniqtuq, *saving the bones for soup.*

a choice of hot Starbucks coffee or chilled Budweiser above ten thousand feet. Back at home in our new post office boxes we'd find state-generated oil bonus checks, glossy flyers, propaganda courtesy of megacorporations eyeing the Arctic, free calendars, welfare and unemployment disbursements, and a stream of slippery Cabela's catalogs.

We had become time travelers, here in caribou country. And the time machine had arrived so suddenly that many of us were still carrying caribou hides for seat cushions.

Here on the "frontier," we had one foot in the past and the other in the future. We were living a fast-forward version of America's mythic history, with grizzly bears and wolves, salmon and caribou, moose and musk oxen, Native tribes and white settlers, hunters and trappers, gold miners and lawmen, old guns and new fast guns, and cowboys on snowmobiles and in ski-planes. The blizzard of change was blowing in too fast for us to have any visibility ahead; there was hardly time to notice that the natural world was changing as fast as we were. The weather was warming—wetter in winter and with thinner ice—and summers were hot and smoky with tundra fires; dwarf trees and brush were blooming into giant versions of their former selves; and old familiar shorelines were disappearing behind tall new willow and alder thickets. Riverbanks and beaches were eroding, melting, caving in; tundra ponds were draining and sprouting fields of grass where twenty years ago otters had swum and loons had dived for fish.

Over the horizon, in courtrooms to the south, the barrage of legal battles deciding the future of our home went on—over jurisdictions, boundaries, subsurface rights, endangered species, traditional land use, subsistence rights, and proposed large-scale extractive development—as politicians and businessmen hatched schemes for northern deepwater ports, airstrips, open-pit mines, floating drilling rigs, and ice roads across caribou calving grounds. Agency personnel struggled to come up with no-fly zones to shield Native hunters from sporthunters, no-waste laws concerning what part of a caribou must be brought in from the field, security guards hired to patrol Native corporation lands, changes to game laws to allow shooting caribou in the river with a .22, and countless other attempts to mitigate the pressures of a digital future colliding with a Stone Age past.

Across Alaska a rural-versus-urban divide flared, closely following the geographic dispersion of Natives and non-Natives. That old high-tension line between Native hunters and white sporthunters grew more acrimonious. But now even the ancient human solution, fighting, was all fouled up in yet another modern dilemma: Who exactly are "we"?

Red Dog Mine in Northwest Alaska is the one of the largest lead and zinc mines in the world.

People were traveling; races were mixing and marrying; Outsiders were moving north; villagers were attending schools and training in cities to the south. Native kids were growing far taller than their parents and imitating life from TV—listening to rap music, wearing baggy pants and team sports gear—and even elders were vacationing in Disneyland. Money was flowing, with new houses and new machinery in every yard, even as the value of furs plummeted and with it the use of traditional clothing. Each wolf killed, and each wolverine, beaver, otter, black bear, goose, duck, and other animal, still carried some prestige, but with each passing season, life became more firmly tied to clocks, workweeks, calendars, store-bought goods, and news and fashions from the big Outside world. Villagers were working stints at Prudhoe Bay, Red Dog Mine, and other work camps, and moving to Anchorage, Fairbanks, Wasilla, and other cities that had been predominantly white—often to return home to hunt as local residents. Early September now brought an influx of caribou hunters from towns, villages, and cities out to the tundra. Hunters laden with new technology, old desires, and an undefined modern need to connect with nature.

Incrementally the monolith *to hunt* had changed, and narrowed. And we had changed. Caribou had forever been important, and it made sense, when grasping for what remained of cultural significance, that they would end up in the crosshairs.

Yet how did our lives become so tilted forward and unrecognizable? How did caribou surpass polar bears, whales, walrus, grizzlies, and even that poster child of land disputes, the wolf, to become our number one political animal? How could something as straightforward as hunting and gathering food and furs become so complicated?

And out there, in spite of this, even tonight, endlessly moving across the tundra, are the caribou.

CHAPTER 15

THE LIVING HORIZONS

Spring is in the air, with crunchy snow, an overcast sky, and a chilly north breeze. Looking down the ridge from our door, I see patches of overflow gray on the white river. The slush has refrozen in the night, and caribou are out there, five or six hundred animals standing on the river, indecisive and awaiting a leader. I've been trying to photograph them this morning, and getting more excited each time another bunch appears out of the willows on the far bank and trots out to join the herd.

I'm sweating, wearing shoepacks and unzipped snowpants, with my gloves, hat, and jacket draped over a sawhorse beside me, cooling down after climbing trees in attempts to get in position to take photographs of the caribou. The big camera I usually carry is in the *qanisaq* now, and instead I'm holding a six-month-old baby. "Look, China," I whisper and point. "Caribou. See them? Those are caribou."

My daughter is wrapped in a blanket knit from *qiviut* (musk-ox wool), and wearing tiny fur mukluks and a muskrat hat. Her eyes are wide in the chilly breeze as she stares around the snow-covered world. Her face registers it all as amazing, although I'm not certain she spots the darker dots of the distant herd. Beside me Stacey grips China's hand, checking to make sure her tiny bare fingers aren't freezing. Having a kid is a new thing for us. Stacey's very attentive, and at the same

Pregnant cows stream out onto the Kobuk River, leading the migration north to the calving grounds.

time accepting of my insistence on continuing our life of tents and traveling the land, seeking food and photographs in the varied seasons of the Arctic. She supports my desire to immerse our daughter in what makes sense to me—nature—albeit cautiously.

"Wow!" Stacey breathes in awe. "Look at them all. Ah! And more coming!"

Her eyes are better than mine, and I hand her our baby and retrieve my binoculars. Together we watch lines of caribou cross the ice to join the swelling herd. "I need to get my camera," I murmur. "I need to go. I have to get pictures."

Stacey wraps her open jacket around China, hugging her in the warm flaps. "We should take her inside, anyway. You go. Good luck. I hope you get some great pictures."

Inside, I stuff rolls of film in my pockets, strap on my pack and pistol. Even as a photographer I'm a hunter-gatherer, and my mind is already outside, shuffling through vantage points to shoot from to capture the best photo to show my miraculous front yard.

When I step out, the herd is still there, nearly a thousand caribou now. I snap a picture, knowing the ice is too white and huge, the caribou too dark and distant, the lighting too flat. I hurry down the hill. Near my family's old igloo, I try another shot. Cottonwoods reach boney limbs into the image. I curse myself for being too softhearted last fall when I was clearing saplings and brush. Trees are growing so fast these last years in the warming weather, so tall and thick—all my old vantage points are overgrown. I jog along the packed trail to the river ice. Thickets of willows poke blurry fingers across my viewfinder. I turn and run back up the hill and frantically buckle on my snowshoes.

Crossing the tundra to a nearby pond, I have to push through dwarf birch—twigs that used to remain knee-high, now reaching higher than my head. I skirt young white spruce trees that have sprouted, grown, and flourished in the last quarter century. I realize again something that is hard to accept: climate change is not waiting; it is not coming, it is here. At the pond, iris seedpods poke out of snowdrifts. This is where Howie first brought Kole and me to hunt geese. New thickets of willows and alders are everywhere, disorienting and confusing. Tall ancient spruce—young seedlings when the Civil War ended and Alaska was purchased—once incredibly slow growing in the former permafrost soil—stand enormous and dark against the sky.

On my old birchwood snowshoes, I jog as fast as I can, tireless and fueled by an intensity stronger even than that I've always felt as a hunter, to photograph these caribou and gather what is necessary to show people how this land would be wrecked by the strip mines, roads, and development that are coming; to convince them to love, respect, and protect it. To me time feels short, and this seems desperately important for the one thing I know the most about: survival.

But protecting this land is tangled up with what I know the least about—people. I have never liked bothering anybody, and similar to Howie, I avoid conflict at all cost. I despise politics, that thing humans seem unable to keep out of their relationship with land. And, as always, I find myself doubting my abilities, doubting my motives, and especially doubting that any "normal" Americans are interested in hearing my wilderness-boy warnings. Most Alaskans think of the land as too large to damage, and Development and Jobs are practically twin religions. In contrast, I'm a tribe of one—not part of any group, too long in the wilds, and ironically taught to fear development and a steady job. I can't help wondering if I'm only trying to protect my own life, hunting and trapping, and fishing.

But what if nature has equal value for everyone, and most people are just too far away and too disconnected from it to see that?

My bushwhacked trail to the river is a hundred yards long, a tunnel cut through willows. The snow has recently been trodden and packed by hundreds of hooves, a white trench marbled with black turds. Clumps of hair cling to branches. My heart pounds, and I tighten my snowshoe bindings and cinch my holster. I check my camera to see how many shots remain on my roll of film, and hurry toward the opening ahead.

The river is a vast white plain stretching two miles east and two miles west. The caribou are a brown-and-white mottled wall in front of me, stretching out of sight up the ice. For long seconds I freeze, almost panicked by this prodigious opportunity and by fear of screwing it up. I'm breathing hard. My viewfinder fogs up. All I see is snow and a dark mass. I snap a photo. The speed is low, the aperture

China Kantner picks berries with her dog Worf along the Akiak River.

wide. *How can the sky be so gray? Am I wasting film and time? Should I be home on the hill with my family, not out here struggling against nature?*

In the herd I spot a cow caribou with a strange, long boney neck. She disappears before I can figure out the reason, and a small dark form catches my attention. Beside it is a third abnormal sighting—a cow with an unusual thicket of large antlers. Is that small animal a black fox? It's too little to be a wolf. I peer through my binoculars. The tiny black creature is a newborn fawn! Suddenly I realize the animal with fancy antlers is a reindeer. I remember hearing the females have much bigger antlers than cow caribou and give birth weeks or a month earlier. I hadn't known their fawns were a different color, too. I glass back and forth, trying to locate the other strange animal. Finally, I get a glimpse and gasp in shock. The poor caribou's neck is grotesque, nearly hairless from jaw to shoulders. Hanging like a loose necklace from its thin bald neck is a wide white belt. A radio collar! I stare, unsettled and as perturbed as if I'd just spotted undeniable proof of aliens here on earth.

Finally, I tuck my binoculars and camera inside my white hunting smock, hunch forward with my arms hanging, and shuffle toward the herd, keeping my face down. Eyes can be frightening, and caribou have had ten thousand years to learn more than a few unpleasant things about two-legged predators.

I'm on the ice, level with the animals. I can no longer judge how many there are, and I curse myself for not staying home where elevation was my friend. But, of course it's never easy to judge what animals will do. *People are even harder*, my mind whispers. For a moment my thoughts race in another direction, wondering what in this world I should teach my daughter. *What will matter in her future? Caribou don't matter to most people on the planet. Or, most people don't believe they do.* Suddenly my thoughts sprint back the way they came: *That has been a problem in the villages, with elders unsure what to teach their offspring in the face of too much change happening too fast. Did my parents have the same dilemma, wondering what to teach Kole and me? Howie would have stuck with teaching us to read ice. And to skin animals, dry meat, shoot, and use hand tools. It was Mama who insisted we speak proper English and learn grammar and math, and study geography and history.*

I reach a slight rise with a buried rock bar. This is where my family set fishnets, summer and fall, especially when I had a dog team and we needed extra food.

China skins a caribou. Stacey carries baby China bundled up in her coat.

I crouch on snow, sweating and shaking from the strain of holding my bent posture and from the knowledge that my pictures will fail to show this astonishing group of caribou. I glance upriver. The dark doorway of our sod house is visible, the roof buried under drifts and smoke floating up from the stovepipe protruding through the snow. Stacey is inside with China, playing with her, cooing and smiling and teaching her new sounds that will be her first words. I wish I could get their attention. I want them to look out the window at what I want to teach my daughter.

Suddenly a wall of animals is moving toward me. The reindeer and her fawn vanish in the rush. Hundreds of caribou sprint in my direction. At the last moment the animals swing sideways and stop. Caribou stare at me. I'm surrounded on three sides. *If only I had a ladder!*

Slowly, I straighten up, clicking the shutter, knowing my shots will show only slices of this caribou horizon. Finally, I stretch my aching back and snap a last image—of a thousand butts now racing away. I lower the camera, awed by the power of one human to frighten so many animals, and dismayed at the disturbance my so-called good intentions have caused. I sigh and absorb the sight and that familiar sound of an entire herd running as one, the river of animals making me dizzy, as if the ice itself is moving.

FOR THE NEXT TWO WEEKS, CARIBOU MIGRATE PAST OUR HOME NIGHT AND DAY, FIFTY thousand or more. The sky stays gray, the weather chilly, and I grow more and more frantic, snowshoeing days and late into the evenings, racing a mile east, across the river, miles north toward the Jades, hurrying up the ridge, west to the ponds, trying different angles and perspectives. Everywhere I travel, caribou move across the land like lines of refugees. The only other mammals I've seen in such concentrations have been humans—in cities—places I fear for the absence of nature, the noise and pollution and cement holding down the land, and that terrible loneliness of a one-creature world.

For now, I'm lost in caribou, hoping my hard work might help shape a future for my daughter, one with huge wild herds intact, and healthy humans who respect animals and the land. Yet on this day, I have no idea that this season and a few more will mark the high point for the Western Arctic herd—probably for my lifetime, maybe forever. I can't see around the corner to the coming century, and I haven't yet realized how fast climate change is accelerating, carrying waves of species decimation. I don't yet recognize how astoundingly rich I've always been to live with the wealth and companionship of thousands of caribou and other wild animals, millions of salmon and other fish, geese and waterfowl and songbirds. How doubly spoiled I've become, using modern inventions to multiply my harvest of the bounty of this incredible land. How blinded by wealth. It will be another twenty years before I begin to glimpse how only deprivation—nothing else—can measure and reveal the value of each treasure in our lives. It will take Septembers with only a handful of caribou on the tundra, and more than one May with an equal lack, for the realization to dawn on me with shock and sadness: I, too, will become one of those old storytellers from my youth, telling tales of times gone by that strain to ring true, that echo like myths in the caverns of time.

maybe already a bit too successful at it. Also, he wasn't from the upriver villages, and for that reason was something of an outsider himself. And he was prone to bragging.

His stories were plenty amazing, but he liked to improve them. And who can say—maybe he did get that wolverine with a screwdriver, and that swimming bull moose with his knife. After all, he could find his way across the Brooks Range in winter without a map, and certainly the number of wolves, bears, and caribou he brought home grew over the seasons to be beyond counting.

Clarence spent countless nights on our floor—more than any other visitor—and days on our bearskin couch, drinking coffee and telling stories, teasing us boys, watching my dad bend sled runners, waiting to see what my mom would pull out of the wood oven. He'd peer over his shoulder out our Visqueen window that flapped in the wind, checking the weather. He never appeared to be in a hurry but often didn't stay long. "Well," he'd announce, put down his cup, and rise. "Thank you much!" Outside, he'd shoulder into his parka, big hands reaching in his pockets for cigarettes. I can still hear the clink of his metal lighter. And out he'd head, disappearing into the land, hunting. That was his life—ceaselessly hunting. That was what he did.

I'm hesitant to say this in these modern overly touchy times, but there was another different thing about him, too: he liked white people. Now that I think of it, I guess that was touchy even back then, because enough folks didn't. Regardless, Clarence found my family and the other stray and strange weird white people who wandered north and showed up along the river to be interesting, and often entertaining in the different and sometimes dumb stuff they did. And they liked him.

My parents and their friends were forever repeating his expressions—even today—and the comical way he put words together. My dad would marvel at how Clarence had made it through a blizzard; Clarence would shrug at the storm. "Ha! C'mon now! Good traveling." Or his description of a bear he shot: "Faaaat. Can't see the meat." Or him describing a winter stuck at home: "Agh. Jus' like jail."

Later, when I started living here with Stacey, Clarence continued to be the one traveler we saw the most often. People were stopping in less, trapping and hunting less, but Clarence never seemed to change. He referred to me as "Tat kid" and her as "Daisy," and continued spending time on the couch, using my slingshot to hunt voles and shrews on the floor, glancing out to check the sky, and rising to disappear into the land. Older, but still relentlessly hunting.

He liked to test me; he'd tease me when I didn't know the channel in the river or I fell through shelf ice; he'd stop in too early in the morning, wanting coffee and conversation, and then ask for something—usually gas. "Agh! Sure need gassss." I can hear that gruff voice, and him saying it. It was true, though. Man, that guy sure always needed another six gallons.

It doesn't matter which of a thousand stories I recall, or what season, what year, Clarence was usually here, or passing by, or had just left. In January 1970, when three Eskimo hunters were murdered a few miles downstream, Clarence was the last one to see them okay before the crime. When Keith Jones's sod igloo next to ours caught fire and burned—there was Clarence on the ice, the first to spot the flames. When that mail plane went down, midwinter, up in a low pass now called Plane Crash, Clarence somehow again was passing by and the first on the scene. He was consistently out traveling the land. And even though this area—Hunt River and Paungaqtaugruk—was his favorite place, if you talked to villagers hundreds of miles away in Huslia, or in Anaktuvuk, or homesteaders up the Ambler River, or whalers up the coast in Point Hope, they'd say similar: Clarence Wood traveled there, too. Hunting.

In the 1970s, when the caribou overwintered here, a few times Grace and Paul Outwater from Kiana spent the night. I remember Grace chiding Paul, telling him not to step on the caribou hides Howie had laid out. Later, when people told Mama that Grace was Clarence's mom, we were surprised and marveled at the idea. Clarence was already larger than life, almost as if he couldn't have a mortal mom, not one who lived only seventy miles downriver.

Clarence was half deaf and seldom talked about his past. I would have liked to know it. My family had heard he was born downriver, or maybe up in camp at the Pah River, and that his dad might have been from the North Slope but had died somewhere in the mountains. One rumor was he died from drinking ice water after eating too much caribou fat. It seemed plausible. We heard Clarence's siblings had died when he was little and only he had survived. That, too, seemed plausible. He was that tough. It made sense.

Alvin Williams and Clarence Wood chat along the riverbank after breakup.

I lost track of how many times I almost shot Clarence. The first time was when I was eleven. I never told anyone. It was May; the first geese had arrived, and I mushed my dog team down to Willow Island across from the mouth of the Hunt, tied the dogs in the willows, and laced on my snowshoes. I sneaked toward where I heard Canada geese hollering. Through brush I saw two dark goose heads moving, disappearing, moving again. The honking was loud, and coming from them. I aimed with my .22 and clicked the safety off. Suddenly I realize it was Clarence's black shaved head in the crosshairs, crouched down behind grass. Beside him was his brother-in-law, Merrill Morena, another good friend of ours, blowing on a goose call.

Another time Clarence wounded a grizzly bear at dusk below our log cache. He'd been drinking, and he left ten minutes later, heading upriver in his boat. I searched for the bear in the dark with a flashlight but couldn't find it in the willows. I came home discouraged and tired, and I laid my shotgun on the table. Stacey woke me up during the night. It was dark, windy, and the dogs were barking like crazy. I went out barefoot. I heard the bear in the brush, coming up the hill. I held the gun level. When it was a few yards away, I aimed and started to squeeze the trigger, but something stopped me. It was Clarence. His motor had broken down, and he'd drifted back downriver. I stretched out a caribou hide for him, and in the morning I heard him pumping the Coleman, softly, the way he did, heating water, wanting coffee too damn early. When I woke up again, he was gone.

Later, bears have pushed on the door and I've woken up mumbling, "Hold on, Clarence! I need to find my glasses and pants." Plenty of times it was him. Once, I woke up my daughter, told her a bear was standing at the door, peering in the window. She sat up, rubbed her eyes, and asked, "Dad, are you sure it's not Clarence?"

"Nah. I already checked."

IT'S CONFUSING, AND NOT EASY TO TRY TO PLACE CLARENCE'S ICONIC FIGURE IN THIS modern world. My own past where hunting was the most important thing feels far away. There are no more hunters like him. Not just no one to take his place, but no place left to take. He was a hunter in the true sense of the word, a predator, a throwback from the old Eskimos who survived on this land. And these passing years have been tough, watching him get old, watching the tidal wave of change, and knowing a time would come when he'd be gone.

In Kotzebue, when he was deafer than ever and wearing glasses, his back giving out, and his stomach hurting, he called me, asked for a jug. I went to the old hotel, had a drink with him. His wife wasn't overly pleased to see the bottle. "When he drinks, it's no party," she mentioned.

"Yeah," I said. "I have noticed that a few times over the years."

Clarence was in pain. "Hard," he growled, grimacing, pushing on his stomach, gripping his lower back. "All my life I push myself. All my life, hard. I never think about give up. Just keep going." His face was serious, in pain, and disgusted. "Agh, my baaack."

The next time I saw him, my family and I had boated to Pipe Spit, east of Kotzebue, and we saw a boat on the Kobuk Lake side of the spit, idling. It came ashore, and I recognized the homemade plywood boat cabin. Clarence was happy to see me—well, not me as much as my brand-new spare prop. "Ha!" he said, gripping it. "You never take it upriver, get ta paint off?"

He gestured for me to put it on for him. He lit a cigarette. "Lotta gas leaking, all right." He held up a wet hose. "This one sure problem me." I glanced over the transom, into his cabin. Two grandkids were up by the windshield, crammed in next to a leaning Mini-14 and other guns in front of blue drums of gas, with gas pooled around the bungs and more tanks in the stern leaking. Quickly, I replaced the prop and fixed his hose. He told me he was turning back to Kotzebue, but when I shoved his bow out, he swung east instead, stubbornly heading across Kobuk Lake at about three miles an hour, smoking another cigarette—hundreds of miles from home, riding a bomb.

When I next ran into him, he was seventy-eight, on the wet spring river trail, below Onion Portage. He jumped off his snowmobile wearing hip boots and a shotgun. It was great to see him, and reassuring, and I told him so. I asked where he was headed. His face was battered, blackened by frostbite. He shrugged that old familiar shrug, scanned the river for any animals, any caribou, wolves, wolverine, geese, anything moving. "Well. No use to stay home."

And then came this fall, with little or no caribou crossing at Onion Portage, and too many boats waiting there. Clarence finally seemed ancient, emaciated, hard faced, stoic, and in pain, yet still hunting, his eyes swinging past people, a predator hungering only for animals. After so many decades with so many caribou, now the tundra was empty. And he faced a double emptiness—his strong body was too weak and worn to go force the land to give up meat.

And now, I know it will be tough for me, to imagine this river and land without Clarence on it. His passing feels like the ice breaking loose, heading out of sight downriver, but this is a much larger change than one of our old seasons coming and going. His passing feels like the arrival of a nameless new season—like the melting ice and warming ground that now threaten winter itself. Clarence was from that old earth, that truly wild land. He was from the old days, and I think

A wolverine runs through the hills near Amakomanak Creek.

for me he'll remain out there with the bears and wolves and wolverine. I mean, if someone told me that after today there would be no more wolverine, they were all dead and gone, I think this is how I'd feel. I'd stand here immersed in memories, cherishing the past, struggling to see a future. While part of me goes on as if he is out there on those old trails, hunting. Because that is the way it was.

FOLLOWING PAGES: *Caribou of the Western Arctic herd aggregate in midsummer to lessen harassment by thriving insects.*

PART IV

THE FINAL SEASON
OF THE YEAR

The air is warm and the sky blue with puffy white clouds reflecting on the calm river. The water level has dropped, and dark smooth muck coats mile after mile of shoreline. Last summer's faded yellow grass is matted under the mud. Willows recently crushed and scraped by heavy icebergs are beginning to turn flinty gray as the silt dries to dust. Driftwood logs, stray branches, and twigs are strewn about, abandoned by floodwaters and tangled with debris and occasional pieces of caribou hide and antlers stranded back in the thickets. Along the riverbanks, snowdrifts lie hidden under dark mud, exposed only where caribou trails have cut through the top layer and left prints in crystalline white snow.

Grass and equisetum poke through the layer of river grime, growing rapidly in the sun-drenched days and nights. Out in the current, young beaver work their way upstream, their backs bleaching reddish brown in the sun as they explore—cautious and anxious, recently booted out of home lodges—and seek territories to start families before winter. Muskrats are doing the same, and pairs of ducks paddle between melting ice and the edges of lakes, leaving lines in the glassy surface, swiveling to scoop floating seeds, quickly moving on, busy and secretive and suddenly waddling ashore to disappear into grasses to carefully concealed nests. The flocks of geese have gone. Only scattered pairs remain, quiet and watchful. Gulls soar above creek mouths, eyeing small fish in the shallows. And arctic terns bounce on fine tapered wings, fishing near the rock bars where they've deposited dark camouflaged eggs in faint depressions in the gravel—effortlessly built nests that they protect fiercely, dive-bombing intruders, guilty and innocent alike.

The spruce trees have pale green tips now, and the birches and poplars have leafed out to form a canopy overhead. Fireweed and bluebells and stinkweed are tall and ready to flower, and in the grass, voles dart down tunnels. Porcupines meander the summer vegetation, feasting on fresh growth and stopping to sit up and scratch and contemplate, and leave behind gray handfuls of shedding winter insulation.

The tundra is a vast green carpet rolling far across the distances, speckled with the white orbs of cotton grass. Aspens and birches stand out against the sky on knolls and prominences, while darker alders dot the tundra and drape distant slopes. Higher up, farther away, dusky blue mountains cradle ravines of snow between rocky gray ridges, and in the rock scree, tiny fragile wildflowers wobble and glitter: blue forget-me-nots and lupines, yellow Arctic poppies and mountain avens, pink mounds of moss campion, wintergreen, and others. In the draws, wild roses and cinquefoils bloom, and across the miles of tundra, Labrador tea and crowberry, blueberry and cranberry, salmonberry and other plants offer their trillion tiny flowers to the sky and wind and insects, asking to be pollinated.

The mountains feel quiet now, aloof and thoughtful, as if they hold a softened beauty and no longer the intensity of the cold and stormy seasons. On the tundra, the summer air is damp and drowsy, buzzing softly, laden with the scent of flowers, sap, buds, and leaves. Tiny tufts of cotton pass high overhead, mixing with the muted conversations of birds and the hum of busy insects. Thunder clouds rise in the heated afternoons, and quick harsh gusts slash trails through the leaves. Rain floods down. By evening the midnight sun shines from the north again, and from the calm current of the silver river comes the slap of a fish jumping, and the patter of loon wings on water. And then in the bright night comes another unmistakable water sound—this one full of meaning and memories—as one last band of caribou wades into the river.

From the far shore, a dozen bulls are swimming in a line. The black velvet on their new antlers is dark against the bleached hair of their backs and faces. Across the water comes the sound of grunts, coughs, and laborious breathing as they struggle against the constriction of botfly larvae in their airways and sinuses. Upriver, from the timber on the bluff, comes the squawk of juvenile ravens, hungry and bored, crapping all over their nest and driving their parents crazy with their demands. The bulls keep their heavy heads pointed north as they swim and roll their eyes to search ahead for danger.

As the animals come ashore, their hooves splash again, making a familiar soft roar. They shake, glance around, sniff the ground, and shake a second time. Wait— are these a different species of caribou? What *are* they? Can caribou be this ugly?

The ungainly animals stand in the grass for a moment, thoughtful and staring around. Their antlers are curved, black and large, softly furred and beautiful. The rest of their bodies seem emaciated, mottled, and strange looking, with clumps of shedding bleached hair and black areas of bare skin. Writhing yellow warbles the

Lupines and other wildflowers dot the landscape.

size of grapes hang from weeping abscesses on their backs and flanks. The skin around their noses and eye sockets is dark, hairless, and scabby from weeks of unrelenting mosquito bites. Their eyes appear huge, sad, and suffering. Suddenly a bull turns and trots into the willows. The others follow, as if they're in a hurry, finally, running late for this migration and realizing that they can no longer avoid it. The willows and alders, grown tall and thick with the changing climate, quickly swallow all sign of their passing.

In the thickets, the animals' tender new antlers tangle in limbs. Rose brambles and dead branches pluck out clumps of their hair. The mosquitoes—now multiplied to a force of nature as intimidating as any blizzard—rise in clouds to ride on the warm wet balding caribou. Before long a line of pale dots files out of the brush onto the tundra. Ahead lie miles of knobby tussocks, uneven and squishy and slippery, tilting and tippy, with boggy holes between each filled with muck, water, and unending hordes of ravenous insects. The small group marches steadily north into the world of tundra, pulled by the draw of some unseen force. These are the last of the last caribou to be seen here until autumn, and tonight with their passing, a loneliness comes on the land.

Winter is gone. Here in its place is this strange season, summer, out of sync with the other seasons of the Arctic, so quiet, with wet water, warm breezes, and ceaseless daylight. And at night, staring north into the sun, you can't help asking: *Could all this light and greenery, all this buzzing and warmth and energetic life fade again to the Darkness, the frozen silence, ice and snow?* The sky says no! These pretty flowers and sweet smells, all these leaves and berries and birdsong, all this sun can never run out! And, yet, the plants and animals are working feverishly, as if every second counts. All the creatures are reproducing as if life might be terribly tenuous, infinitely valuable, and the survival of their offspring the only thing that truly matters.

ON THE OPEN TUNDRA FAR TO THE NORTH, THE PREGNANT COWS HAVE REACHED THE CALV-ing grounds. They are following food, as the land begins greening up, and have switched from eating lichen to plants of the season—now cotton grass that has sprouted from the rounded heads of a billion tussocks, the buds and stems providing the caribou with nutrients they need.

As the first grasses begin to flower in balls of cotton—soon to paint the tundra white as summer snow—the caribou spread into smaller and smaller groups, each animal an individual, none a permanent member of any group, as is the way with this migratory animal. The expectant cows have begun to feed less and spend time staring around nervously, ready to flee a mile or more to a new location if they become suspicious of danger.

In the first days of June over the course of a week, tens of thousands give birth, totally exposed, in sight of the entire world, out on the treeless plains north of the last northern mountains. The intrepid mothers quickly lick and nuzzle and nudge their newborns, facing them and bobbing their heads to get their calves up and moving, and to establish connection and imprint each other's smell for vital future identification. At the age of one hour, the calves are on their feet and beginning to nurse. Within a few more hours, their mothers stare around, glance at the horizon, step forward and clear their throats, and introduce their children to the first rule of being caribou: "Ert! Come on. Get your stuff. Time to go. Ert!"

And off they go, twelve or thirteen pounds, hours old, and moving out into large life. The calves are wobbly, tilting on gangly legs, but already nibbling their first bites of cotton grass and other green plants. Less than a day old, they are ready to begin traversing terrain that most adult humans would find daunting beyond imagination—mile after mile of knobby tussocks, scree slopes, snowdrifts, cold swift rivers banked with thickets hiding bears and wolves. The astounding abilities of the newborns are matched only by the unrelenting migratory urge of their parents; like the mass birthing, this instinct to move is a way of dealing with predators, but also an adaptation to protect the extremely sensitive Arctic tundra from damage by overgrazing and in that way maximizing the caribou population.

Within a few days, cows with similar-aged newborns form nursery groups. Soon larger and larger bands of caribou are on the move. The cows drop their antlers within a week after giving birth—the last of the herd to do so—and again lead, now with tiny tawny calves close at their sides. Non-maternal cows begin to join them, and again the bulls grudgingly bring up the rear. The bulls' new antlers are growing still, and are large and curved and covered in dark gray velvet. The animals' hooves, too, have altered with the snow-free season; the long hard black shell has worn away, and the fleshy pad in the center has grown thicker.

As the herd moves, along the way calves are taken by grizzlies, wolves, golden eagles, and other predators, with the highest toll happening in the first week or two after birth. If the caribou young survive that period, they've often doubled their weight and are able to outrun a grizzly bear, and in another week or two can be traveling fifty miles a day across hard terrain.

As the sun moves toward the solstice, the mosquitoes grow worse and drive more caribou to join the migration. Lines stretch for miles like white threads across the green tundra, some traveling east, most moving west, seeking high ground and the accompanying breezes to keep insects at bay, traversing slopes and broad valleys, crossing rivers, climbing mountains, flowing through an endless maze of thousands of ancient trails that vein the land. Along the way, bleating fills the air; tiny calves call to their mothers—*ert! ert! ert!*—and their mothers call back, and stop, and urge them on. The cows allow their young to nurse for only short durations, less than minute, to maintain close contact and strengthen the bond between them. When the quick meal is over, they continue onward. *Ert! Ert!* They persuade their offspring to leap down cliffs, plunge through thickets, splash into rivers, and fight terrifying currents. Calves are swept downriver; bears and wolves disrupt entire nursery groups; frantic mothers run or swim back against the tide of oncoming caribou, calling, searching for their offspring. Lost calves run from cow to cow, attempting to nurse, only to be quickly rebuffed and threatened for their mistakes.

Without natural fear of predators, the small calves will be swallowed up if not quickly located. Their mothers know this and call out, agitated, anxious to find their lost kids and to keep them close and safe. And still, in those first weeks, as many as 40 percent or more of these intrepid young caribou will be eaten, lost and starved, or swept away in river crossings.

UNDER RELENTLESS SUN AND TWENTY-FOUR-HOUR DAYLIGHT, THE CARIBOU MOVE ACROSS the tundra, traveling steadily. July brings increased heat and harassment from even more mosquitoes, horseflies, and now warble and botflies, too. This maddening horde of insects drives the caribou to congregate and seek high mountains and any remaining snow. In huge masses they take advantage of sheer concentration and constant movement to reduce the suffering of each individual animal. In this annual gathering, nearly the entire population of the Western Arctic herd (and likewise the Porcupine and other large migratory herds) form into groups of as many as fifty thousand to one hundred thousand animals, or more, gigantic aggregations of a kind no longer seen in the rest of North America, and rare now throughout the world.

For a few weeks, the caribou remain in close proximity. As the insects begin to lessen and the weather cools, the animals disperse into smaller groups, finally free to

Wildflowers grow near rock formations along the Kokolik River.

focus on food and fattening up before their next major migration. They feed on sedges, willow leaves, mushrooms, and late-flowering tundra plants, and the cows use this time to recover from the strain of the past few months. The bulls meanwhile are putting on fat for the coming rut. The herd's next major journey, beginning in mid-August, will be the fall migration to the wintering grounds—windblown tundra to the south—although, in recent decades with changes in weather, increasing vegetation, decreasing ice formation, and alteration of the timing of the seasons, the caribou have changed the timing of their traditional migrations and grown more unpredictable in their movements.

August in the Arctic in recent years has been swelteringly hot; Septembers have broken record highs—far above freezing and with no snow or ice—and temperatures in early October have reached sixty degrees *above* zero. The longer warm season causes the tundra to green earlier and reduces forage quality for caribou later in the summer. As the permafrost under the surface vegetation thaws, the composition of the formerly frozen ground changes, and across the Arctic plant communities are in flux: microbes in the soil, frozen for millennia, are undergoing rapid transformation, altering the underground environment as much as or more than the visible changes above ground. Trees, brush, and formerly dwarf Arctic plants are now growing as if on steroids. Boreal forests are expanding rapidly.

Thermokarst lakes—with water once sealed in by permafrost banks—are undergoing sudden catastrophic drainage; methane and sequestered carbon dioxide is leaking from the melting ground; and streams and rivers are remaining unfrozen much longer and later in the season. The ice freezing later—and to less depth in winter—is bringing countless changes in turbidity, erosion, and habitat permanency for aquatic creatures and at the same time physically inhibiting the movement of animals, including caribou. Humans, too, are facing growing danger from changing conditions, and drownings through thin ice and open water have increased.

Winter itself, the defining season of the Arctic, has turned into a messy, uncertain, and tenuous thing, arriving a month or more behind schedule—too warm and too wet—and often now a surreal season of rain, slush, and weak ice, all taking place in that old familiar Darkness. The Arctic Ocean and Chukchi Sea have become completely ice-free for extended periods, even in winter, causing increases in evaporation, precipitation, and winter storm frequency and severity. Rain in winter—extremely rare in the past—has greatly increased and often coats the tundra in ice or layers of ice, affecting rodent populations, furbearers and other animals, and making it hard or impossible for caribou and other ungulates to access lichen and preferred vegetation.

As the land, soil, water, and air warms, the caribou's food—and the timing of that food supply—has become unpredictable. Summer fires in the forests and tundra have also increased, destroying lichens that can take half a century to regrow. Meanwhile because of the increasing humidity and warmth, more insects are surviving and new ones are arriving, causing the animals to face additional and unfamiliar hardships.

Summer, with all its heat, bright light, and abundance, is not offering hope. Weather impacts everything a caribou does, decides their every day, and too many changes are coming too rapidly. With eons of experience enduring the harshest conditions, caribou are now caught in the midst of these cataclysmic changes with no time to evolve to face these new threats. They can only adapt, or perish. They have no choice as the earth under their hooves transforms into a different planet.

FOLLOWING PAGES: *Stretching for miles, lines of caribou head west across the Kokolik River after leaving the calving grounds. A cow gives birth on the Utukok Uplands. Adult bulls again bring up the rear as the herd migrates away from the Utukok.*

Antlers and bones are all that remain of an adult bull that died while crossing the Great Kobuk Sand Dunes. A young wolf pup peers out from a patch of fireweed.

FOLLOWING PAGES: *As the ice-free season lengthens and permafrost melts, the Arctic coastline erodes at a faster pace. Botfly larvae grow in caribou's airways for nearly a year before hatching. Forest and tundra fires are becoming more frequent. A storm surge raises the sea level in Kotzebue and floods a dog racer's yard.*

Caribou cross the quarter-mile-wide Kobuk River during breakup.

CHAPTER 17

BOUNTY HUNTER

On a sunny afternoon in May, the sun feels extra intense, burning my face and melting the snow. Everything is slushy and settling as I run on my snowshoes, a mile downriver from our igloo, to try to intercept a group of caribou. I'm not after a photograph or meat. Today I have an actual all-American job, working for the State of Alaska.

The sky is blue, and songbirds and waterfowl are calling. The temperature is rising toward sixty degrees above zero, and the air is rich with the smells of the land thawing and the sounds and excitement of breakup. Along both shores, moats of open water are dark and frightening. Ripples and whirlpools wink in the sunlight. The water is coming up fast, and stray ice pans sail west in the current like floating white packages. Needle ice tinkles and bobs, hinting that the main ice might move today for the first time.

The solid ice stretches down the middle of the river, wide and smooth and still snow covered. A string of dark dots is out there, black against the white—caribou crossing from the south, with more coming out of the willows and swimming the narrow moat to climb up onto the ice. The herd is angling this way, toward where I'm crouched behind a clump of willows, peering out to check their progress.

I'm shirtless, my body slung with gear. Skinning knives press into my waistline, my hip boots are cinched to my belt, and across my shoulders and neck the straps of a machete, a camera, binoculars, a backpack, and my AR-15 rifle crisscross and gouge my bare skin. A faded yellow rope crosses my chest, poking painful nylon splinters into my skin. The rope is attached to an orange plastic sled I'm dragging in case I end up needing to haul meat home. In my pack, a gray metal zero-to-two-hundred-pound scale clanks against my kidney. The snow is soft and heavy, hard going, but I'm in the best shape of my life, tireless, fearless, focused, and intense. I'm happy, too, at home out in nature, and helping science—whatever that really means. Because of my perpetual hunter-gatherer worldview, I love physically tough work with a clear purpose, and now finally I'm a biologist's assistant of sorts, and ironically a bounty hunter, too. I'm using expertise I've honed all my life to contribute to caribou research, and make money. I've been hired by ADFG. My job description: shoot caribou calves. The pay, a hundred dollars apiece.

There's more to the story, obviously, and I'm trying not to let my mountain of misgivings darken my day. There's a surprising coincidence here, too, that I haven't allowed myself to think much about—how I'm repeating a portion of history that Howie lived thirty-five years ago, when during the Cold War he shot caribou for samples for analysis of radiation levels on the North Slope, and chased down newborn calves to place ear tags on them. Now I'm contributing to yet another idea hatched by scientists to study caribou—one I found appalling last fall when I was first offered this job.

Over the years, slowly, I'm coming to the surprising realization that big game biologists are taught to protect and respect numbers, not necessarily individual animals. This is the opposite of the way I was raised, with animals as not just my food and daily focus but also my most consistent companions on the land, and this perspective feels wrong to me on many levels, as if too much respect for numbers might lead to a blindness to the plight—or a lack of empathy for—the animals themselves. I've spent the bulk of my life staying home, right here in sight of Paungaqtaugruk, the bluff where I was born. Staying home, and hunting animals and birds and fish to survive. In my forays out to college—paid for in dollars earned from furs and salmon—I took only a handful of classes in biology, oceanography, and atmospheric science before switching my major to writing and photography.

But there is another less admirable force at work here, too; I was informed that if I declined this job, biologists from Fairbanks would come north with helicopters and guns and do it anyway. The Western Arctic herd is huge, over 450,000 animals—considered too large to be sustainable by some—and game managers are looking for a way to forecast a decline that they believe is coming. Like most hunters here would be, I was anxious to keep helicopters far away, and also pleased, proud, and thrilled to be singled out to help scientists "help" caribou. Now, I've got myself caught in what feels like a trap: still wanting to be useful, struggling not to be judgmental of the project and biologists whom I've never met, while at the same time feeling there's something tragically wrong with shooting caribou calves to predict a crash of the herd.

The lead cow has just reached the edge of the ice. Fifty yards of water separate her from shore. She stops and stares across the current, straight at me. Slowly, I drop my eyes. I hold my breath and wait. Quick movements will give a predator away, and at times it seems as if staring at an animal can deliver a message that a hunter doesn't want to send—an invisible sensation of unease. Incrementally, I crouch lower. The sun and the glare off the snow burn on my arms and forehead. Lately, on the AM radio news, there has been talk about the ozone layer thinning in the Arctic, and on this warm, bright day I feel as if the *Death Star* is searing my skin. Kneeling, I smile at my hands in tense excitement and reach to cinch my snowshoe bindings tighter. The cow has decided not to swim. I watch as she turns and trots downriver. Within a few minutes, the leaders disappear out of sight around the bend. The string of animals methodically follows. I need to move.

Hidden by willows, I thrash along as fast as I can. Trees catch my snowshoes, branches poke at my face, scratch my arms, and hook my rifle barrel. I don't mind the scrapes. My skin is a map of scars. And I like being tanned, or darkened with frostbite scabs—anything to help mask one of the largest hurdles of my life here: appearing white. Sparrows flit and dart out of the way. A rabbit streaks off into the brush. *Lucky guy, able to turn brown to match his surroundings.*

Bulls with dark new antler growth and non-maternal cows and yearlings bring up the rear of the spring migration.

Behind me the plastic sled tangles, bouncing along, loud and irritating. When my snowshoes finally poke out of a thicket, I stop and quickly grip willows—below me a sheer cutbank drops eight feet to current. The main ice sheet is close here, a quarter mile wide, flat, open, and inviting. The reflection off the snow is intense, cooking my face. *Oh, man, I want to be out there!* But the moat is twenty feet wide and the current dangerous and swift, dark and disappearing under the ice. And the whole river of ice may start moving at any moment.

Caribou pass steadily. I can see the knobs of warbles under their skin and the black nubs of new antlers beginning to grow on the larger bulls. Sweat runs down the inside of my sunglasses. *What should I do?* I'll have to wait for a different group. Day after day I've been waiting, trying harder, waiting more. Unlike Howie, I don't appreciate waiting. I feel as if I've already done more than my share of it in my life. I prefer action to thought; I want intensity. I want to bust my ass, hard and dangerous and every day, for a true purpose. That's partially why I agreed to do this job. Now I'm feeling a moral dilemma, a pressure to get these caribou, or the helicopters will come ratcheting overhead. A character from a book I read as a kid comes to mind, as it often does—Dick Summers in *The Way West*, a mountain man turned scout and guide for wagon trains of settlers crossing the frontier—and his internal struggle knowing his wilderness expertise was being used to help destroy what he loved.

I glance downstream along the cutbank. Fifty yards farther, the main ice almost touches the gravel shore. I twist my feet out of my snowshoe bindings, pull my rifle forward over my shoulder in one movement, and crouch behind the clump of willows.

Seth, this is not a good idea. You can't get out there to retrieve what you shoot. I'm breathing too hard, anyway. I need to wait. I examine animals with my rifle scope—just to see if I can pick out calves. The biologists want me to weigh and note fat content on ten calves born last June. That was eleven months ago; the calves have grown, most no longer trail their moms, and now are not always easy to tell apart from yearlings. The yearlings have longer faces—that's the best place to start.

The caribou are moving in single file. All the young animals seem to have long noses today. Suddenly the line stops. Caribou swing to face me. The pregnant cows have hard brown antlers still, and they stare suspiciously at the shore. There are only a few bulls in the herd, and they stand looking, too, but not vigilantly, as if they want no part in expending the energy it takes to make dangerous decisions. *Who can blame them?*

I lower my rifle gently, raise my binoculars. The group begins moving again, heading downriver toward the mouth of the Hunt. Beaten trails there mark where thousands have come ashore. In front of me I count calves, one-two-three-four. My hands pick up the rifle. It's crazy to think of gutting and skinning animals out on that ice. Risky to even attempt to get out there, and then somehow get back ashore before the whole river breaks loose. Experience and a subliminal feeling tell me the ice is going to move today. When it does, as far as I can see in both directions will be one massive moving sidewalk. A quarter-mile-wide smashing crashing deadly white sidewalk. With a billion gallons of icy-cold water swirling underneath.

I swing the rifle, examining caribou. They're so amazing. Such brave animals, moving steadily down that ice, facing hardship and danger and death every day, every step, with unwavering determination. My chest fills with love for the caribou, and all the animals, and for this big beautiful home we share. I think of Howie, falling in love with the Arctic, and I feel a rush of gratitude for my parents for raising me here. *Who would I have been in Cleveland or Toledo, Ohio?*

I decide to leave these creatures in peace. It's best that I snowshoe back home and watch for more caribou from the hill. I grimace, realizing I'm a bit too philosophical for this job. Alvin, or Clarence, would be single-minded. Howie taught Kole and me never to kill something for no reason, not to waste a life—especially a beautiful young calf that just survived its first winter. For a second I realize the imbalance in my thoughts, how as usual I'm more worried about not wasting an animal I can't retrieve than about myself drowning. Howie definitely didn't teach that. But like so many other things in our world that don't add up, I am stuck as a permanent hunter-gatherer wherever I go in this modern society. Me, with my cheek comfortable against the black plastic stock of this rifle, aiming, considering, and falling into the hunting zone. Accepting risk as part of life. Here I am, alive, and capable, and completely in this moment. Like these caribou, death has always been a part of my life. *Boom!* The gun instantly chambers a new round. A wave of sound smacks into an animal. A calf crumples to the ice.

Caribou divide smoothly, sprinting, fanning out across the ice. They halt suddenly, trying to locate the invisible danger. The calf is dark, dead on the glaring white snow. *Boom!* A second calf drops, farther out on the ice. The AR might have been designed specifically for this job, it is that perfect. As always I'm sharply aware of the difference between an auto-loading weapon and a bolt-action hunting rifle, and without looking up, I swing, aim, fire. A third calf is down. Caribou run and mill, and small groups stand in confusion.

I swing the rifle, hold for an instant on the head of a fourth. The face seems long. Abruptly I raise the barrel, forcing myself to stop, recognizing something

Millions of moving ice needles fill the air with tinkling sounds and create cold fog over the river. A young caribou is marooned on the flowing ice.

I've felt before when hunting caribou. Something I've felt when hunting geese and wolves, too—but never moose, bears, or beaver—an impulse that must have to do with a group of animals finally being within range, after much effort, much hunger and struggle and repeated searches. What I recognize is an impulse that takes over and makes it hard to stop dropping one animal after another. A semiautomatic rifle makes the impulse worse—faster, easier, uninterrupted—and harder to stop. I wonder, do wolves and bears and other animals that kill for a living ever feel this? Maybe it's a result of extreme expended effort, sudden opportunity, distilled desire, and of course possessing an awesome power to take lives. It's hard to explain. I don't quite know what it is. I only know that the force is real and that Howie spoke of it, too.

I unload the chamber, ram the clip in, and scramble to gather my ejected brass. Caribou plod west again. Across the ice near the far shore I see more coming, swimming the moat, and climbing onto the ice sheet. The air is quiet. Suddenly from below me comes a roar. Water cascades off a surfacing ice pan. Startled, I stare at the main ice and line up a dark spot with a tree on the far bank. The river is white, still, unmoving, exactly where it's been since October.

In the next breath I spring into motion, yanking off my camera and binoculars, looping the straps on a branch to leave beside my rifle and snowshoes. I toss the sled down, grip the base of a willow, and leap over the bank. In the air I twist, hang for an instant, and drop to gravel shore.

The air is warm below the cutbank, full of the smells of water, thawing dirt, and decaying leaves. I grab the sled and run along the bank downstream until the moat ends. Ice pans are jammed between the main ice and shore, and dark current gurgles and flows under the white ice. Suddenly I remember that I need saplings to make a tripod to hang my scale, to check the ice, and to save myself if I plunge through. I squirm free of my pack, claw and cling to roots as I climb back up the bank. In a thicket, I yank Howie's old machete out of his moose-hide sheath and cut three large willows. Scratches on my arms are bleeding; I fling the poles over the bank, leap down.

Out on the glaring white ice, the line of caribou is skirting the dead animals, accepting the loss. I recognize something I've seen many times before: how visible dead often make caribou more at ease. Countless times caribou have approached me in open sight, even surrounded me, as I bend over a dead animal to gut it. Part of this is curiosity, and often I've "called" them in by waving a white cloth, but there's more. Maybe when danger is everywhere—and predation is normal—a sort of relief comes from knowing a price has already been paid, and from being able to clearly see the threat.

I toss two poles to the main ice, fling the sled after them, and pole-vault to an ice pan. Current boils black in openings under my feet. As kids, Kole and I were taught the dangers of ice and open water at this time of year. But I'm an adult now, alone, and modern times are different along this river; hunting still matters, but risk has a new addictive and uncertain place in it. Dying while hunting has an undefined cost. These days no one will starve if I drown, and sadly the reverse of that—"what is my worth?"—has become an uneasy question for many of us young men to fathom. I have certainly thought, many times, that my daughter would be better off without me to lead her down my anomalous path through life.

I jump to a second pan. All around me the ice tinkles and jostles and creaks. The whole river is ready to go. I leap again. I'm on the main ice! It's as if I've arrived on a new continent, and I grab up my gear and sprint across the slushy snow. At the first calf, I check that it's dead and quickly swivel to check the shore. No movement yet.

"Thank you, little miss caribou. Sorry it was you." I yank a knife, slash cuts at the ankles and wrists for handles, stuff a rope through, cinch the legs tight together, and tie a knot. The caribou is shedding, and my hands are instantly coated in hair. With another rope, I form the willows into a tripod, hang the scale, and lift the calf. I'm sweating. Biting on a pencil, I bend to read the weight—forty-eight pounds. I lower the calf, rush to untie the knot. I wring my hands clean of hair, pull at the belly skin and quickly cut open the body cavity, and then free the brisket on one side and fold it back. With bloody fingers, I grip a yellow Rite in the Rain notebook and jot my information: *Calf #4, F, 48lb, min intest fat, medium heart and brisket fat, no back fat. Moderate warble larvae. Liver 1 cyst.*

A stray warble lies on the snow, big and gross and yellow. For a second I glance around. I've read how Natives ate them, but I've never tried one. Now seems to be a good time. My hand puts it in my mouth. It's soft, wet, and faintly salty, and I swallow quickly. I switch knives and remove the boney lower legs and head, flipping the head upside down to take out the tongue. I heave the carcass in the sled, snatch up the heart and tongue, tripod and pack, and race toward the second animal.

Later in their migration, the caribou face hardship and danger as the ice grows unpredictable.

I'm weighing the third calf—*#6, M, 68lb*—when a slow swell of sound rises like a gust of wind arriving. Needle ice shifts. Small piles sigh and cascade. I glance at a tall dead willow on the bank. *I'm moving!*

Behind me the line of caribou halts. Caribou stand staring at the shore passing us. I drop to my knees, split the calf open, peel out the entrails. Already the whole ice sheet is picking up speed. There's no time to cut off the legs or head—no time to count warbles or remove the tongue. The pans I used for a bridge are floating free. I lash two caribou on the sled, swing another over my shoulders, and stumble toward the shore edge of the ice. For a moment, I freeze. *Which way should I go? Downriver!* I have to go west, even though it's straight away from home, my snowshoes, and rifle.

A whole section of my world has broken lose. Everything has changed with the ground under me moving. The ride is exhilarating but comes with no guarantee of survival. I've ridden ice many times before but always with a kayak or a small boat, rope, and a life preserver, and usually with Kole or Alvin accompanying me. Now I'm out here alone, with a load that's too heavy and hundreds of caribou traveling beside me. A sudden thought flashes in my head: this whole experience, including being hired to shoot calves for science, is a metaphor for how my old hunter-gatherer life is sliding toward some untethered and uncertain new reality.

Time has vanished. I feel an awareness of death, like a new quality of light, but without any fear. Everything has ceased to exist except now, here. I'm simply a capsule of life, bursting with adrenaline, panting and shirtless and bloody to the elbows, traversing the truest kind of danger—the opposite of anything I've found in cities—where I'm allowed, prepared, and able to use anything and everything in my power to survive.

The main ice is ramming into the cutbank, coming to pieces under the stress. Long sections crack and shear off. Sheets tilt as they collide with shore. Ice pans rise, piling on top of each other, roaring, pulverizing, rising higher, and falling in heaps. A crack opens at my feet. I step across, yank the sled after me.

Which way should I go? I need to get ashore before I'm cut off by moats, before I pass the flowing wide mouth of the Hunt. The weight of the caribou slows my decisions. I can't leave the meat. I can't waste it. It has to come with me. Finally, I reach the edge and stare longingly across the water. I drop the caribou off my shoulders and kneel and hack at the waist, twisting and cutting, yanking as hard as I can until it separates into two halves. I fling one, then the other, onto a floating ice pan.

A fissure opens behind my boots. The line lances out hundreds of feet across the ice. Dark water fills the space. The air is full of the roar of moving ice. I leap to the main ice, change my mind, and leap back across the widening fissure. I tug the rope. The orange plastic sled bridges the opening. I'm on a raft of ice, fifty feet long and being shoved along by the momentum of the main ice sheet. The lower end is swiveling toward shore. I run to the middle. My sled surges forward, suddenly lighter. A caribou has fallen off. I run back, snatch it up over my shoulder. Under me dark needles squeak and collapse. I plunge down into water. Instinctively, I throw my body forward. The weight of the little caribou on my back pins my chest against solid ice. I thrash, claw at the rough surface, haul myself out. My fingers burn and bleed. I tug the sled across floating needle ice. My legs are cold. Quickly, I dump water out of the sled and my boots.

My ice raft shudders. The shore edge tilts, lifting as it gouges into gravel at the base of the cutbank, rumbling as it smashes into loud beautiful white mounds. Cracks are forming everywhere, and I leap over one after another, run up the angled ice ramp, yank my sled close, and jump. I land in slush and jumbled ice. My pack whacks my back. I struggle to keep my footing under the weight of the caribou. Everywhere needles are tinkling, crashing like heaped glass. I crawl over loose ice, onto sand. *I'm on land!*

I scramble up the shifting ice to avoid being crushed against the bank. Frantically, I toss gear up into the willows. With fumbling hands, I tie a rope to the caribou, and around my waist, pull myself up the bank, and lift the caribou after me. *I made it?*

I turn and shade my eyes. I still have to retrieve the other caribou. A small gray form catches my eye thirty yards upstream. The pan with my meat on board is being shoved along in a jumble of other pans. My view is great from up here, and I pause to watch the huge spectacle below and to quickly jot down information on the third calf. When the pan is below me, I lower myself to a pile of ice and pole-vault from pan to pan. I plunge through twice, but each time I'm able to use my stick to fall across and crawl back up onto the ice. When I reach the halves of the calf, I turn and fling them, one pan at a time, as I work my way to shore.

Finally, it's time to start the hard trip back to my gun and the rest of my gear. And then the longer, slower slog home. Once I get there I'll stoke the barrel stove, rush to make coffee and fry a heart, and wolf it down. I won't rest, but instead will begin skinning the animals, cutting hindquarter meat into strips to hang and dry.

Caribou rest on the main ice in the middle of the Kobuk River.

I'll dig a big hole and bury the rest in the snow. I'll be tired, but at the same time tireless. Around me gray jays and sparrows, the birches and alders, lichens and a first few big slow mosquitoes, and so many other lives, will be busy with their own endeavors. Toward midnight, I'll slide a shoulder roast into the oven and pick some melting-out cranberries to make myself a pie.

I've been alone here for weeks and have taken to talking to the trees and chickadees and myself—inside and outside of my head—and while I work, my mind will travel its own tough terrain, a tiny invisible me in there endlessly navigating thickets of thoughts, leaping over old composting convictions, stopping to peer at new ideas, sinking waist-deep in doubt—questioning what I am working so hard to accomplish, what true thing can I really do to protect this amazing wilderness, and how will I ever find something as pure as hunting to help sustain it? In the bright cool night songbirds and waterfowl will be calling, and out in the river the ice will have jammed, caribou will have resumed crossing, and sometime before the sun starts again to climb the sky, I'll crawl into bed and call it a day.

CHAPTER 18

BLACK FRIDAY

Caribou were scarce this late summer and fall, again delaying their migration, and local hunters were in a frenzy to find them, boating the Kobuk and Noatak Rivers, burning gas, staring at the shores and at their phones and GPSs, and finding few animals. Through political maneuvering, Native groups had gotten Outside hunters restricted from federal land, although, ironically—aggravating to locals on the ground—those hunters who flew north to locate the herds had success landing with small planes on sandbars along waterways (recently ruled by the US Supreme Court to be under state jurisdiction).

During freezeup the herd suddenly began to move. The caribou followed the coast—not the healthiest choice of trails—and before long a two-hundred-mile-long vein of animals stretched from Cape Thompson to Kivalina, south to Cape Krusenstern, and on all the way to Buckland and beyond. In front of Kotzebue dots were visible on the white ice, plodding past both ends of town. People could peer out of landing Alaska Airlines passenger jets, or stand on Front Street, or even in line at the post office, and watch out those big windows as thousands of caribou passed.

There one day, mailing a flat-rate package and eyeing caribou working their way across jumbled ice, I stopped to chat with Johnny Walker,

A calf stays for days near the head and gutpile of its mother, shot by hunters.

the son of Howie's old friend Nelson. Johnny is an entrepreneur of sorts, like his dad was, and I always enjoy talking with him. He told me about an old rusted dozer he was repairing to plow willows off one of his Native allotments, and about a new hunting venture he had started. Under state law a person can't hunt caribou the same day they've been airborne—so to work around this he'd bought a reindeer herd, which are legal only for Natives to own, and a helicopter, and he could now fly sporthunters directly from the Alaska Airlines terminal in Nome to his private herd. There his clients had him shoot a video of them "hunting caribou," and then while his assistants took care of the trophy and cape, he quickly choppered clients back to catch the next jet south. I stood gripping my mail, shaking my head. *You can't make this stuff up!*

The ice was still thin, not entirely safe, and I confined myself to town a bit longer. I know how to check ice, and find thin ice educational and exciting, even nostalgic. What I was protecting from drowning was my sanity.

Finally, I took my camera and walked down the shore past the end of the airstrip, where a crowd of trucks, snowmobiles, and all-terrain vehicles was lined up. Nearby small groups of men brandished an array of semiautomatic rifles. I grimaced but reminded myself that I carried an AR-15 for twenty years straight, nearly everywhere I went. Hell, I even took it to the University of Alaska with me, leaned it against the wall in my dorm room: 607 Bartlett Hall. Suddenly I pictured my roommate, Mark, borrowing it to shoot holes in a TV out on the old gold dredge tailings.

Offshore, a cow with a calf at her flank was leading a line of animals across uneven ice. The caribou were nervous as they moved perpendicular to the men. Onshore the testosterone was thick in the air, a stressful veil of competing egos and something intense that felt unpleasant, like hate, or getting even with those reviled airplane hunters. It reminded me of being at Onion Portage in recent falls, watching as hunters raced each other in boats at high speed into swimming herds of caribou, guns blazing as they came—a practice locally referred to as "combat hunting."

I nodded to a few guys I knew. None of us appeared to be starving. Everyone was clothed, slung with, and riding on gear that cost hundreds and even thousands of dollars—including me with my Nikon. Old traditional needs allegedly had gathered us there, but new unacknowledged hungers, I think, loomed larger still. I snapped a few photos and retreated. Howie had always been uncomfortable

A group of animals attempts to come off the ice near the airport at Kotzebue.

to see animals killed as a way to connect with nature or harvest manhood, and I have worked to stay true to that training.

I went home, waited to hunt, and struggled to write. My book was taking too long, too many years, and feeling useless. Who cared anymore about caribou lives and struggles? Didn't most people consider, say, the stock market infinitely more important? Onion Portage was on Wikipedia now, and hunts and treks and tours to caribou lands were listed on Google. Climate change was in the process of melting the world around us—Northwest Alaska is ground zero for global warming—and the absence of local, regional, state, and national response to the crisis was causing a crisis in me. I felt sapped of my unquenchable energy. I didn't feel as tireless as a wolverine anymore; I felt scattered and restless, and strangely short-tempered. I stared at my sentences, listlessly deleted pages, dithered day after day over comma placement. My mind visualized an asteroid plummeting toward Earth, with me up there frantically carrying heavy buckets of dirt from the front side to the back, in attempt to delay impact—by what, a millisecond?

Daily, I distracted myself by answering emails, texting Alvin and other hunter friends, and staring into the windows of that magic store: Amazon.com. In the afternoons when the news came on the local AM radio station, I gratefully ceased writing and cleared a place at the table to eat lunch alone on a piece of cardboard: dried caribou, the last of the sweet carrots and turnips Stacey and I grew this summer, berries, and bowhead *muktuk* from a friend in Point Hope. The home harvest tasted good and felt familiar. Most days I eat about that same food while listening to the news, and afterward I heat water to make coffee. Howie grows my coffee on the farm in Hawaii where he and Mama now live, and that connection comforts me, too.

The Alaska news mentioned something else besides caribou that I had been spending too much time thinking about: the Ambler Road, a project proposed by multinational mining conglomerates to have the state pay to bulldoze a billion-dollar road through the Brooks Range to Ambler. The road would allow those companies access to our homeland, one of the last largest intact wilderness ecosystems in the world, to turn the mountains upside down for copper, silver, gold, and other minerals. The agencies, corporation leaders, and politicians had held hundreds of meetings about it, for decades, claiming to want public testimony, claiming to be weighing local concerns over caribou and other subsistence resources.

I had to switch it off. Corporations aren't designed to care about caribou. Those heavy rocks Tommy Douglas searched for long ago in the hills, that's what they want. And suddenly I wondered what I did with that lump of bornite he gave me years ago.

Pacing around my papers, I missed walking on the fall tundra and new ice and knew I should be helping break trail to the future. But how? I believe in nature, and the way forward involves people. I missed my daughter, down in California beginning her junior year at Stanford—intent on saving the world too, and soon to have a degree in environmental studies. In September, at Paungaqtaugruk she'd gotten her first caribou and shared meat with elders, an old custom. The weather had been unseasonably warm and I'd waited to hunt, and ended up with no caribou. Now it was October and the bulls were rutting, which would exacerbate climate change's hidden tolls: the delay in the migration had again caused local hunters to miss out on harvesting bulls in September; they in turn would alter their patterns

A hunter and his son chase a cow and calf on the ice.

and pursue pregnant cows on the tundra with snowmobiles. And all the while, biologists kept reporting an unexplained high cow mortality rate these last few years.

I knew the story well. A modern hunter on a snowmobile is no wolf, nor grizzly—as far as stress on the animals—and gets the whole group running, as it effortlessly pursues the animals mile after mile. And that machine will often return the next day, and the next. Soon even a distant drone will make caribou flee in terror. Without enough stress-free time, the animals get run-down, unhealthy, skinny. And with machine chasing comes herd shooting, and wounded caribou that don't get salvaged. Small-caliber Mini-14s and AR-15s in careless hands make matters worse, and caribou hyped on adrenaline and fear don't always show a hit from a small 55-grain bullet. If they do limp or fall, they're often shot multiple times from the rear, hit in areas that ruin meat in the hindquarters, back, and shoulders. The targets—females, many still with calves—are smaller and thinner and lighter than the bulls, and pregnant, too. Most have little fat, and once shot, the meat often freezes quickly without aging properly. When factoring in bullet damage, wasted meat, wounded animals, fetuses destroyed, orphaned calves, and stress on the entire group, the per-pound per-animal harvest in winter is far lower and the cost to the herd far higher than when a conscientious hunter carefully harvests a large bull before the rut.

IN NOVEMBER I COULDN'T TAKE MY SELF-SEQUESTERING ANY LONGER, AND I PACKED MY camera and .243 rifle, fired up my old snowgo, and headed out to see how the warm weather was affecting the ice and how the caribou were faring. It was the day after Thanksgiving, Black Friday.

The tundra was plastered white from the last storm. Up on the ridge behind Kotzebue, I stopped and glassed the sprawl of town and the frozen ocean. Down near the airstrip, I could see where Mamie Beaver's tiny shack used to be and a few feet away the old Drift Inn, now housing Native corporation offices. Around me, the snow was trampled with caribou trails. Below the ridge, the ice was alive with activity: groups of migrating caribou milled off Front Street, and a mile farther out a large herd was headed for town. Snowmobiles zoomed here and there. Dead caribou were visible on the ice, circled by the tiny black blips of ravens. To the north toward Lockhart Point, half a dozen machines were parked with people hunched nearby, gutting their kills. As I watched, three more snowmobiles streaked away from town, traveling like comets, snow billowing behind as they raced toward the herd. I stared in dismay as they scattered the migrating animals into frenzied fleeing bunches. It was like a video game, except with a beautiful orange sunset out toward Siberia, and the bobbing yellow headlights, rapid gunfire, and flowing dark lines of terrified running animals.

I set up my tripod and camera, peered through the viewfinder. I looked away and kicked snow. "Ramboing" caribou—modern local slang for chasing down animals—has never felt like hunting to me. Shooting at animals is simple; hunting well and respectfully is incredibly difficult. I guess I'm old. I learned the old ways from Howie and other old-timers, back in a different time. I learned walking the tundra, or snowshoeing hard behind hills and along draws, crouching down, sneaking closer—until finally peering over a rise, and then lying there in the snow, staring—intently watching as the caribou feed, deciphering which animals are rutting, or limping, or nursing, or fat, then aiming, carefully, and shooting. And walking back for a sled.

Up on the ridge, I sighed and clicked my camera off. I sat staring out across the ice. It was so flat and big and beautiful. I wished I had some heroes, one or two people self-aware enough to be bigger than their smallest hungers, even just one person to follow through this battlefield between humans and all the other creatures trying to live on this planet. I'd heard too much rhetoric about respecting animals, including that often-repeated line about letting the leaders of the herd pass unmolested, but who actually does those things? I'd seen too much waste, too many wounded animals—entrails and fetuses shot out of running caribou, caribou with flapping broken legs, abandoned dead caribou—camouflaged behind both kinds of hunting, sport and subsistence alike. I'd seen *National Geographic* photographers, wildlife biologists, and others display equally damaging and egocentric behavior. *But me, how am I different?*

By then the sun was gone. The sky glowed orange. My camera was cold and I put it away. My gun was untouched in the scabbard. The sky was big and blue, the ice gray white, and out toward Sisualik lines of animals were still coming. I wanted no photo of this. I needed meat—or believed I did—but of course I could eat sheefish, or rabbits, or beaver, or maybe find a moose, or countless other store-bought foods. Like every hunter in this region, I was allowed five caribou per day, 365 days a year—but now I had no stomach to add to the herd's hardships. It felt like Black Friday had arrived here in the land of caribou and caribou hunters.

In the following days, acquaintances reported dead and wounded caribou scattered around Kotzebue and other villages. Worse were the calves—orphaned and wandering the town—a calf behind the Alaska Commercial grocery store, one huddled at the airport, three at the softball field, and more near the hospital and in peoples' yards. Out on the tundra, I found stray adult caribou, but again couldn't make myself take out my rifle.

One young bull stood beside the trail as I passed. In the low sun my shadow crossed his face. I stopped. He didn't move, or run, or turn away. The poor caribou stared at me as if I were an alien. I felt heavy, weighed down by all we've lost, and I wished we could stop long enough to stand in silence, listen to the land, and realize that what we are killing is not caribou but ourselves.

CHAPTER 19

AN OLD GRAY TABLE ON THE LAND

On my floor mixed in with my cameras and winter gear and guns and heaps of notes is a plastic CD case. The cover has a photograph of Bob Uhl, old and weathered, wearing a big white parka with a wolf ruff. I remember taking the photo on that spring day at Sisualik; the snow was bright and glaring off the sea ice, Bob was seventy-five or so—old as the hills, I thought—and gripping a shovel, digging out his buried wall tent. Now I'm over fifty and he would have been close to a hundred. Time feels different to me, the years getting squishy, easy to compress, and going by much quicker than they once did.

I miss him and his wife, Carrie. They were local icons in this region, for living out along the coast year-round since the 1940s. Bob, a transplanted Californian, came north as a young man in the wake of World War II and met Carrie, an Iñupiaq Eskimo. Together, they spent more than half a century living in a wall tent in spring, summer, and fall on the shore at Sisualik, and in a small cabin in the trees a few miles away during midwinters.

So many times over the last decade, I've wanted to go back in time—to ask Bob a question about caribou sinew, or calfskins, or what he

A river otter waits for me to eat the blackfish it shared.

thinks of my theories about the crash of the herd. Or I want to ask him about pink salmon, and snowy owls, ice conditions, overflow, and humans, too. Which way is *that* herd going to go?

I miss Bob's encyclopedic knowledge of the land. After a lifetime out here, his experience was vast, and he had mental capabilities to match. He also dearly loved to delve deeper and deeper into the details of individual species. It feels weird to admit this—especially for someone who doesn't care for computers—but his passing leaves me feeling as if the "google" of nature vanished from my life. *Why didn't I ask a thousand more questions while I had the opportunity? And record every answer?*

The battered CD holds twelve years of his journals, filled with details of daily life, although only tidbits compared to what he held in his head. To tell the truth, I've only read a few pages and heard short passages on the radio. Something about reading about Bob Uhl's life on a screen gives me pause. Maybe this is the time, though. More than information about caribou, I could use some lessons in his acceptance of the world, his generous perspective, his ability to witness bad behavior and yet continue believing in the goodness of people.

I just came from a meeting of a local subsistence advisory committee, my second in two days. My proposals—one to limit the top speed at which people can chase caribou with snowmobiles, and the other to lower the total number of animals each hunter is allowed (currently 1,825 caribou per year)—got ridiculed and I was accused of being racist. Meanwhile another person's proposal, to solve the problem of orphaned calves by making it legal for everyone to shoot them, sailed through. It felt so ludicrous, it was hard to know where to turn.

Bob had once told me about his experience with subsistence meetings, boards, and advisory councils. He had made trips across the ice to town one winter to attend such meetings. It wasn't for him, he said. He mentioned something about local politics, and town hunters who didn't spend much time out on the land. He chuckled and was good-natured about it, though, and I remember being frustrated—me young, wanting action, and despising empty rhetoric; in awe of his knowledge of the natural world; and impatient with his chosen role as humble observer.

A bull urges an orphaned calf to its feet as the caribou migrate south.

I'd first met the Uhls when I was nine, and over the years my family and I regularly made trips across the ice or water to visit Bob and Carrie. Their home at Sisualik was on a harsh, windy, wave-battered spit of land. Beside their tent, the remains of various past activities poked through the chaff of beach grass, a tired old dogsled, rusted snowmobile chassis, a bent harpoon, gray forlorn dog stakes, a white sun-bleached killer whale skull. Nearby were weathered fish racks for hanging *ugruk* meat to dry. Dug in the ground was a moss-covered *siġluaq* (cold storage), and draped under tarps was a small freezer, powered a few hours every other day by a portable Honda generator.

When it wasn't windy, raining, or too buggy, Bob and Carrie ate outside at an ancient wooden table, the legs buried in storm-flung pebbles on the crest of the beach grass overlooking the ocean. Waves glinted in the sun, gulls and other birds passed overhead, and the sky sprawled away in all directions. The table was crowded with food harvested from the land, much of it dripping with fat: pot-roasted caribou, baked trout, king salmon collars, the ubiquitous seal oil, and preserved in the seal oil was black meat, braided seal intestines, *tukkaiyuk* (sea lovage) greens, and occasionally *masru* (Eskimo potatoes). There was beluga *muktuk* on a plate, and sometimes *kauk* (boiled walrus skin) if a hunter had gotten a walrus or one had floated in. There were *aqpiks* (salmonberries), *paunġaqs* (crowberries), *quaġaq* (fermented sourdock), and Royal Cream crackers, canned milk, sugar, and tea bags.

Along the beach, out of the path of passing Honda four-wheelers, Bob kept a very long pole—actually many old gray poles wired together—and if the wind, tide, and season were right, he would use it to push a sack of rocks tied to a fishnet out through the surf, before we started eating.

At the table, a few yards away the loud rhythmic surf swept the broad pebble beach, my blue plywood boat lifted in the swell at anchor, and beyond it the line of floats marked Bob's small net stretched out in the waves, and while we ate we watched for fish splashing and corks bobbing as the meal went on and stories were told and news exchanged.

IN THE PHOTO ON THE CD COVER, HIS EYES ARE DEEPLY SQUINTED. THERE'S A WRY GRIN on his lips. He didn't have any teeth and was bald under his billed hat with unruly wisps of white hair. I can't help smiling, remembering how when you asked Bob a question—about caribou, starfish, south wind, or salmonberries, or any question concerning the natural world—his blue eyes would twinkle and he'd chuckle and not answer. Not right away. Because, well, really you didn't even know what you were asking, and he would have to explain that, too. Which was fine with him. He had the time, and patience, and he greatly welcomed the conversation. Living out on the land comes with a loneliness and a desire to share stories, and share the knowledge and food and warmth that you have spent your long days gathering. And I guess there's a lesson there, too: you can develop a great love and appreciation for humans when you don't see them very often.

Bob, when do the bulls start getting fat again in spring? I can see him now, hunched in the entrance of their *siġḷuaq* reaching his huge hand into a wooden keg of seal oil to scoop out blubber and black meat for lunch, and him pausing, absorbing my question. *Ha. Well, Seth. Hold on! First, I need to ask, how hungry are you? Have you eaten?*

Bob would offer food and something hot to drink, and make sure I took time to talk with Carrie and sign her visitor book. It was a process. Hospitality was the law of the land, and as Bob gathered more food to put out, he would be gathering his thoughts to answer the question.

That kettle just boiled. There's tea bags in the box there. The trout was fresh yesterday. Where were you yesterday? Ha. You asked about bull caribou. Well. Ha. Obviously that depends what you mean by bulls. Older than three years, one would assume. It depends what you mean by spring. And what you mean by fat! Fat is the only bank account for many of our fellow creatures—you know that, Seth. The bull caribou must have his fat for winter . . .

Bob went on, explaining the minute details of fat placement on an animal and how the caribou's bones—especially the bulls'—begin to grow thinner from the inside, in February, which makes the opening larger for *patiq*, which translates to more fat. This change in the bones is less visible on cows, he explained. He chuckled and mentioned how hunters nowadays, since the recent arrival of snowmobiles, choose to pursue cows because of their more-visible back fat; they have forgotten the traditional value of bone fat, and—"*Arii*, Bob!" Carrie would shout. She'd slap his shoulder fondly. "Stop talking! Too much about caribou!" Her pink flyswatter would strike the table, nailing a fly buzzing around the berries and seal oil.

We'd all laugh, but I would be still staring at Bob, awaiting even one more detail. Bob would grin and apologize for his tendency to go on. And he may stop talking about caribou and bones and fat—or not—but you knew there was a lifetime of information stashed in his head about this single subject, and the same for nearly every species in sight for miles around that old gray table.

Bob taught me to bury caribou meat in the winter to keep it from freezing and to allow it to age; he told me of the first moose immigrating here in 1948 and how hunters shot it and determined it to be good eating. He described the arrival of the first snowmobiles, and the disappearance of dogs; he expressed his great admiration for *Eriophorum angustifolium*, cotton grass, and elucidated on that plant's importance to not only caribou but also many other different species of birds and animals; he also told stories of herders who could track a single reindeer traveling with a herd of caribou, and how in the old days when people needed sheefish in late

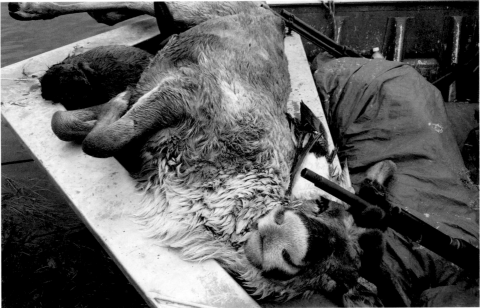

Bob Uhl shovels out his and Carrie's tent to prepare to move to the Sisualik coast in spring. Village hunters launch their boats and head out to hunt as soon as the ice breaks up.

winter, diminutive Mamie Beaver could chip a hole down through the ice faster than the largest and strongest men. She was tough and tireless, but also could mysteriously read the snow covering the ice to find thinner spots. Bob told of walking north to harvest meat and caribou skins, and of pursuing foxes on snowshoes, and how there used to be only a few clumps of willows on this coast, and if you knew them all you could use a shortcut on the fox because those willows were where it would hide. He explained how to scrape the ice to fool a seal, and to blow air into the seal's lungs after you shot it so it would float behind your kayak.

Bob said things that surprised and confused me too: he wasn't concerned about people losing traditional knowledge and old ways of doing things. *Well, Seth. People will relearn it, if and when they need it.* I've forgotten most of what he told me, but I do recall one thing he insisted I learn. *Seth, animals are individuals. Each one is an individual. Like us. You need to remember that if you're planning to learn about our fellow creatures.*

He told me this after my lifelong belief in the integrity of science had been badly damaged. I'd already seen musk oxen blinded by radio collars flopped forward over their horns, and wolverine with their windpipes cut by the hard sharp edges of similar collars. Then I was hired by federal biologists to take photographs while they attempted to collar Dall sheep. What I witnessed at the time was the chasing of animals with helicopters, and arrogant inept men—"fly-in biologists" who had never seen this population of animals before—holding the sheep captive, literally torturing them, prying teeth out of their jaws with screwdrivers, poop out of their butts, ripping hair from their necks to locate veins they should have known how to find with the pressure of their fingers—and in the end holding the animals too long in too much fear, until they developed capture myopathy and were paralyzed and dying. When I expressed concern, I was promptly relieved of duty, and a chopper was called in to remove me from the mountains.

Dismayed and searching for a hard-hitting hero, I told Bob what I'd witnessed. Instead of channeling my anger, however, he offered food, friendship, and observations. It took me years—maybe until today—to realize those quiet actions were, in fact, heroic, and maybe what I needed most. Bob listened and chuckled and agreed that he wasn't comfortable with collars on his fellow creatures. He chewed at his

CHAPTER 20

MAKING *PANIQTUQ*

Tonight, I'm cutting meat to dry, making *paniqtuq*. The woodstove is warm and the windows black and flexing in the wind. On the AM radio an announcer is talking about a new sickness called a coronavirus. It sounds as if nature is throwing something at us. A grain of sand. Or something smaller actually, smaller than a particle of dust. I wish I could say I'm surprised. It won't be good if it gets to the villages, especially not for the elders. I remember old-timers telling me of the 1918 flu—the igloos full of dead people, and searchers cutting holes in the sod roofs to look down in, to listen for kids crying among the dead and pull out a baby or a toddler and leave the rest. That pandemic killed both of my great-grandparents on my father's side. Little like that has happened in our lives, and the past is easy to forget.

Anyway, tonight seems like a good time to dry caribou meat. Those elders in the ground, Mabel Thomas, Minnie Gray, Mamie Beaver, Clarence Wood, Tommy Douglas, Bob and Carrie Uhl, and others, they would nod—serious and wise—and agree.

I make *paniqtuq* by freezing a hindquarter and then thawing it enough so I can shave the meat thin, like shaving a popsicle. Not everyone freezes it first, but soft meat is wiggly and harder to cut thin and trim off sinew and layers of tissue between muscles. That is not

Packaging fresh paniqtuq *for my family's lunches* (Photo by Sarah Betcher)

Caribou march across the ice, accepting life and death along the trail.

necessary either, but it does make the meat dry faster and makes it easier to chew. A big thin, sharpened knife or an *ulu* works well.

I shave each slice, trim off sinew and any glands, and keep an eye out for small tapioca-like specks: tapeworm cysts. We used to not worry about eating glands, but they can concentrate an array of modern toxins. Tapeworm cysts are degraded by freezing, and by drying too, and hypothetically humans are not one of their hosts, but it is more reassuring to avoid them.

Mama used only salt to season the strips. These days I use a mixture of sea salt, which my family gathers, and pepper. Mostly, I want to be able to taste the caribou. I layer the slices in a chilled pan, season each layer, and then cut more and add more layers. I work fast because it's easier to separate the slices to hang before they thaw into a gooey undefinable pile. Mama used cotton string, hung up near the roof poles of our old igloo. I remember yellow-brown strips of sinew stuffed up there, too, and dust from woodsmoke, and Howie's white ptarmigan wingtips. In recent years, I've made a bunch of drying racks with wooden Xs like a clothing drying rack, and also double strings to keep the meat from sticking to itself. Lately I've liked my simple wide wooden rectangles better—strung with herring fishnet to lay the strips on. People I know also use store-bought dehydrators, but I don't have one.

Put cardboard or newspaper on the floor to catch the drips. If you're near any sort of electricity, place a small fan nearby to speed up the drying. If the meat takes too long to dry, it can go bad, or end up tasting like your house. I cut and hang *paniqtuq* in the evening, and the next morning while drinking coffee, I break the strips loose and heap them in a pile to finish with the fan lying on top. It's hard not to eat the driest strips right then, they're so fresh and good.

When I was a kid, my family ate dried caribou and fish for lunch. We ate it with seal oil or bear fat, and later in the winter with *quaq* fish or *uilyak*. In the summer we ate blocks of pemmican. We made those with caribou *paniqtuq* pounded to pieces or run through our hand-crank meat grinder, placed in a pan with dried currants and cranberries, and poured over with hot rendered caribou fat. When the fat cooled and before it hardened completely, we cut the panful into bars and later wrapped them in cloth to store away. It all took time. It was great eating, though. And we had plenty of time.

There's something about food from the land that your hands gathered or hunted or grew, similar to a skin you've tanned—something beyond physical sustenance. Maybe you value the food and furs more because you know how much work they took, every step of the way. Maybe that appreciation is in part gratitude, which nourishes us too.

Anyway, tonight feels like a good time to make *paniqtuq*.

WHITE WALKERS

The light is falling as I head down the ridge toward our old igloo to glass the tundra for caribou. The mountains are dark against the northern sky. In the warm air, an insistent cloud of gnats buzzes around my face. I see no animals, and in the growing darkness I feel disoriented by the tall spruce, alders, dwarf birches, and willows. I wonder, would I even be able to spot caribou coming through all this new brush?

In the morning, Stacey and China and I take berry bowls and Howie's old .270, and walk to an old favorite cranberry spot, a ridge my dad named the Far Birch Knoll half a century ago. Along the way the blueberries are big and chalky blue—no frost yet—and the bright tundra sweeps in all directions, back the way we came, to the Jade Mountains, and far up the Nuna Valley. We see no sign of caribou. The land feels empty without their presence.

On the knoll, the cranberry plants are lost under gigantic Labrador tea bushes. The lake below the hill, once blue water, is now a mat of floating green grass. Two swans are out there, and ducks feed nearby. Along the shore, an old beaver lodge I used to trap from is still in use, although the grassy slope that leads down to it has grown in with alders.

Stacey and China take a nap in the sun, and I leave the rifle with them while I search thickets for a bear den I spotted a few years ago

A group of large bulls crosses the tundra near the Jade Mountains.

and sky, gathering food—it's fall after all, and hunting and gathering is what I've always done. I don't want to struggle with unruly words or untangle thoughts mired in misgiving. I wonder why simple words feel so daunting. What is causing this feeling? Fatalism? Hopelessness? Desire to avoid conflict? A breakdown in trust in our democracy? And what kind of love for this land leaves no time for truly defending it?

I miss the old hunters stopping in: Clarence Wood's face at the door, at any hour, black, frostbitten, needing gas and wanting coffee—sitting awhile, joking, and telling stories of hunting, and then rising stiffly, groaning, aching from the miles but unable not to stare hungrily out at the land and travel on. And Alvin, from the dawn of my memory, Alvin was always out there.

LATE IN THE MONTH I PEEL SPRUCE POLES, CUT BRUSH, AND WATCH FOR SNOW. THERE IS none, and the ptarmigan and rabbits have begun turning white, preparing for winter; they stand out, terribly exposed on the still-brown land. In the evenings I stay outside as the twilight falls. The branches of the birches and aspens on the hill are bare, jagged forks reaching like delicate black lightning up into the night sky. Across the river, the timber on the far riverbank is dark. I listen in vain for the great horned owls. They were there, every night, for fifty years. Why have they gone?

On a day when the water temperature has dropped to near freezing, I begin to pack my family's old wooden boat. I need to return to the coast before ice. But the wind rises and rain comes, cold and gray, stormy. One morning snow has painted the mountaintops. The waves are still too rough for me to travel. I wait, walking the ridge, relentlessly watching the tundra and river for caribou.

In the evening, I decide to try to make it to the village. I board up the cache, shed, *siġluaq*, and *qanisaq*. With my binocs I glass north for a last time. In the fierce gusts I blink away tears, and stare. A mile-wide army of caribou is marching across the wet brown tundra. *White walkers!* my mind whispers. How and from where did thousands of caribou appear? Everywhere the tundra is dotted white. As I watch, more crest the horizon. A wall of animals is coming toward me. Is this a dream? No, this is life.

ACKNOWLEDGMENTS

Over the years I've been slow to notice things I can learn from animals and other living creatures. Sled dogs were the first to illuminate that for me. I remember my pups playing tug-of-war with a caribou hide, and then suddenly fighting over it. My best friend Alvin and I laughed, for decades, about his dog Cooey—actually his dad's old lead dog that we unceremoniously yanked out of retirement and harnessed and hitched to Alvin's sled—because of the way she'd trudge along, slow and glaring back over her shoulder, and when we kids weren't paying attention, she'd lunge sideways, whip the sled around, and sprint for home with us and the team in tow. It took me too many years to write this book and along the way I often reminded myself of Cooey. In this case, Kate Rogers, editor in chief of Mountaineers Books, was riding the runners and didn't allow any turning back. I'm not exaggerating; this book would not have happened without her.

Back when I was a trapper, and steaming wood to build sleds, Stacey Glaser came into my life; later, as my wife, she steadfastly supported the thousand dilemmas of a struggling artist, and accepted my adding writing and photography into a subsistence lifestyle—all in a modern world. My parents, Howard and Erna, offered that too—amazing acceptance of every trail their sons ever chose.

My daughter, China, countless times skinned animals and helped with caribou meat, offered an opinion on a photo, or waited in the cold while Dad was messing with his tripod, or, more recently, found information I needed to make progress on this project.

My agent, Sydelle Kramer, has never wavered, nor has my opinion of her. She encouraged me to take on this project, guided me, and wouldn't take a penny. The Rasmuson Foundation provided support, and the Whiting Foundation more support and sheer belief in my work. It's hard to know where that comes from.

In the good times and the bad, editor Emily White answers—and answers again when I need advice to decipher her previous advice. Don Rearden does that too, even more often, as does my brother, Kole, who somehow talks me into not using boiling water or an axe to fix this old computer.

Thank you, Helen Cherullo, Laura Shauger, Ali Shaw, Tess Day, Jen Grable, Laura Grange, Darryl Booker, and the rest of the folks at Mountaineers Books for the countless behind-the-scenes-and-screens details it takes to make a book like this happen. Thanks to Alice Bailey, fine editor, who read more than one version, and Dan O'Neill—yet again—for sharing pitch-perfect Alaskan insights, and Kimberly King Jones, who appeared like a sparkling Star Trek technician beamed down precisely when I needed a caribou biologist who wasn't afraid of politics.

Nick Jans, longtime friend, combed my final manuscript; Larry Kaplan and Hannah Loon helped with Iñupiaq language; Peter Lent with North Slope details; Geoff Carroll—biologist, dog musher, hero of mine—with perspective; Anne Beaulaurier spotted my seagulls and other mistakes, and spent many hours staring at photos with me; Christie Osborn and Alex Hansen answered stray questions; Ole and Sasha Wik shared photos and stories; Molly McCammon sent enthusiasm and a packet of ideas; Lew Pagel let me stay in his cabin as I was finishing this book; and Peggy Shoemaker and Joe Usibelli did the same when I was starting it.

Thank you, Eric Sieh, for friendship and knowledge, and being someone to talk to about this tilted life, and always for your unbelievable piloting skills. I won't forget those night flights far into caribou country—especially since I don't usually even like to leave the ground! Thank you, Tom Campion, for your support for a couple of those trips; I expected to be intimidated by you but instead was too impressed to have time for that.

Thanks to Don and Mary Williams, for visits and meals, coffee, journals, photos, and laughter, love, and friendship down through the decades; Mary always with more coffee and another loaf of bread and a bag of *paniqtuk* before I headed back out. How could I do it without you both?

Finally, thank you to the Iñupiaq elders, many dead and gone, for sharing stories and lessons, knowledge of the land and to the dedicated scientists in Alaska, and around the world. Anthropologist Ernest "Tiger" Burch devoted his life to understanding the history of caribou and people in northern Alaska. His book *Caribou Herds of Northwest Alaska, 1850–2000* was invaluable to my efforts. *Caribou and*

A lynx swims to the far shore of the Kobuk River.

the North: A Shared Future by Monte Hummel and Justina C. Ray was helpful, especially in understanding boreal caribou, and field reports of the Alaska Department of Fish and Game provided useful background reading over the years I worked on this book. Many caribou biologists have spent their careers contributing to our knowledge of these animals. Hunters and non-hunters alike have benefitted from the knowledge you gathered, sifted, and shared. I have largely left your names out because of politics—not disrespect. I recognize the risk you may face from speaking freely, and that some of you were unsure of what might become of your hard-earned information in my hands. That's okay. You know who you are. And that goes for the people who told me stories of the land too—so much important knowledge packed into so many stories, from old and young people alike. Keep telling me more please.

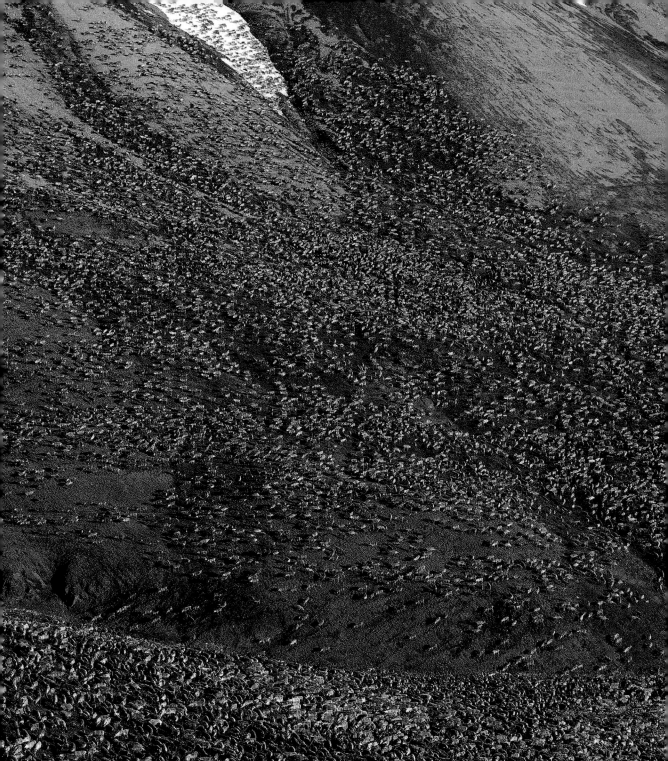

IÑUPIAQ GLOSSARY

Adding an -s for plural nouns (listed in parentheses) reflects modern usage.

akutuq Eskimo ice cream

alappaa it is cold, to get cold

aŋatkuq shaman

aqpik (aqpiks) salmonberry

arii expression of hurt, disappointment

ichuun skin scraper tool

itchaurat lacy stomach fat

kauk boiled walrus skin

masru roots, Eskimo potato

muktuk whale skin with blubber

niqipiaq Eskimo food

paniqtuq dried meat

patiq marrow

pauŋaq (pauŋaqs) crowberry

qaatchiaq (qaatchiaqs) caribou sleeping skins

qanisaq entryway, storm shed

qiviut musk-ox wool

quaġaq fermented sourdock

quaq frozen meat or fish

siġḻuaq cold storage in the ground

siksrik (siksriks) ground squirrel

siksrikpuk marmot

tiktaaliq mudshark, burbot

tukkaiyuk sea lovage

tuttu caribou

ugruk bearded seal

uilyak frozen, fermented, raw fish

ulu (ulut) curved Iñupiaq knife

umiaq (umiat) skin boat

Caribou crowd together on mountain slopes to limit bug exposure.

ABOUT THE AUTHOR

Born and raised in northern Alaska and homeschooled on the land, Seth Kantner has worked as a trapper, commercial fisherman, gardener, adjunct professor, and wilderness guide. He has spoken about climate change and life in Alaska throughout the US and in Canada and has taught writing and photography in Alaska and the Lower 48.

Kantner received a Whiting Award for his debut novel *Ordinary Wolves*, which also won the Milkweed National Fiction Prize, the Pacific Northwest Booksellers Award, and *Publishers Weekly* Best Books Award, among other honors. His memoir, *Shopping for Porcupine*, was chosen as a book of the year by the University of Alaska. Kantner is also the author of a children's book, *Pup and Pokey*, and a collection of essays, *Swallowed by the Great Land*. The recipient of two Rasmuson Foundation awards, he also received a Whiting Creative Nonfiction Grant for *A Thousand Trails Home* as a work in progress. His writing and wildlife photographs have appeared in the *New York Times*, *Smithsonian*, *Reader's Digest*, and *Outside*, as well as many other publications and anthologies. He has been a columnist for *Orion Magazine*, the *Anchorage Daily News*, and *Alaska Dispatch News*.

Kantner attended the University of Alaska and University of Montana, where he earned a BA in journalism. A commercial fisherman for more than four decades, he divides his home between Paungaqtaugruk and Kotzebue, where his family lives.

MOUNTAINEERS BOOKS is a leading publisher of mountaineering literature and guides—including our flagship title, Mountaineering: The Freedom of the Hills—as well as adventure narratives, natural history, and general outdoor recreation. Through our two imprints, Skipstone and Braided River, we also publish titles on sustainability and conservation. We are committed to supporting the environmental and educational goals of our organization by providing expert information on human-powered adventure, sustainable practices at home and on the trail, and preservation of wilderness.

The Mountaineers, founded in 1906, is a 501(c)(3) nonprofit outdoor recreation and conservation organization whose mission is to enrich lives and communities by helping people "explore, conserve, learn about, and enjoy the lands and waters of the Pacific Northwest and beyond." One of the largest such organizations in the United States, it sponsors classes and year-round outdoor activities throughout the Pacific Northwest, including climbing, hiking, backcountry skiing, snowshoeing, camping, kayaking, sailing, and more. The Mountaineers also supports its mission through its publishing division, Mountaineers Books, and promotes environmental education and citizen engagement. For more information, visit The Mountaineers Program Center, 7700 Sand Point Way NE, Seattle, WA 98115-3996; phone 206-521-6001; www.mountaineers.org; or email info@mountaineers.org.

Our publications are made possible through the generosity of donors and through sales of 700 titles on outdoor recreation, sustainable lifestyle, and conservation. To donate, purchase books, or learn more, visit us online:

MOUNTAINEERS BOOKS
1001 SW Klickitat Way, Suite 201 • Seattle, WA 98134
800-553-4453 • mbooks@mountaineersbooks.org • www.mountaineersbooks.org

An independent nonprofit publisher since 1960

FOLLOWING PAGES: Page 315: *A young porcupine feeds along the recently frozen shore of a pond.* Page 316: *Caribou struggle to cross the final few yards to shore.* Page 317: *Pregnant cows (with hard antlers) lead a line of animals through slush.* Page 318: *Caribou migrate across the ice of inner Kotzebue Sound.* Page 319: *Crossing lake ice after a rain in late fall to retrieve my traps.* Page 320: *After the rut, a bull rests and ruminates before continuing down the long trail.*